Topics in Applied Physics

Volume 120

Topics in Applied Physics is a well-established series of review books, each of which presents a comprehensive survey of a selected topic within the domain of applied physics. Since 1973 it has served a broad readership across academia and industry, providing both newcomers and seasoned scholars easy but comprehensive access to the state of the art of a number of diverse research topics.

Edited and written by leading international scientists, each volume contains high-quality review contributions, extending from an introduction to the subject right up to the frontiers of contemporary research.

Topics in Applied Physics strives to provide its readership with a diverse and interdisciplinary collection of some of the most current topics across the full spectrum of applied physics research, including but not limited to:

- Quantum computation and information
- Photonics, optoelectronics and device physics
- Nanoscale science and technology
- Ultrafast physics
- Microscopy and advanced imaging
- Biomaterials and biophysics
- Liquids and soft matter
- Materials for energy
- Geophysics
- Computational physics and numerical methods
- Interdisciplinary physics and engineering

We welcome any suggestions for topics coming from the community of applied physicists, no matter what the field, and encourage prospective book editors to approach us with ideas. Potential authors who wish to submit a book proposal should contact Zach Evenson, Publishing Editor:

zachary.evenson@springer.com

Topics in Applied Physics is included in Web of Science and indexed by Scopus.

Bernard Doudin · Michael Coey · Andrejs Cēbers
Editors

Magnetic Microhydrodynamics

An Emerging Research Field

 Springer

Editors
Bernard Doudin
Institute of Physics and Chemistry
of Materials of Strasbourg-CNRS
University of Strasbourg
Strasbourg, France

Michael Coey
School of Physics
Centre for Research on Adaptive
Nanostructures and Nanodevices
Trinity College Dublin
Dublin, Ireland

Andrejs Cēbers
MMML Lab
Department of Physics
Faculty of Physics
Mathematics and Optometry
University of Latvia
Riga, Latvia

ISSN 0303-4216   ISSN 1437-0859   (electronic)
Topics in Applied Physics
ISBN 978-3-031-58375-9   ISBN 978-3-031-58376-6   (eBook)
https://doi.org/10.1007/978-3-031-58376-6

This Springer imprint is published by the registered company Springer Nature Switzerland AG
The registered company address is: Gewerbestrasse 11, 6330 Cham, Switzerland

If disposing of this product, please recycle the paper.

To
Ausma, Donatella and Wong May,
for their 140 years of love and patience

Preface

This volume in the Springer *Topics in Applied Physics Series* results from a research programme initiated by Bernard Doudin and Thomas Hermans in Strasbourg, inspired by early work of Michael Coey at Trinity College Dublin. The idea was to combine magnetic and microfluidic circuits to create new ways to transport matter. The University of Strasbourg International Advanced Studies programme (USIAS) in synergy with the Chaire Gutenberg initiative supported stays of Michael Coey in Strasbourg and the development of these novel ideas. With a key input from Peter Dunne, a new collaborative dimension was added through the European funding of the Marie Sklodowska-Curie Innovative Training Network (ITN) on Magnetics and Microhydrodynamics, which ran from May 2018 until June 2022. Some of the outcomes form the core of this volume.

The ITNs aim to train a new generation of creative, entrepreneurial and innovative early-stage researchers; in our case, the successful training of 14 young Ph.D.s by the nine academic and industrial beneficiaries;

Raheel Ahmad	Gunatilake	Vahid Nasirimarekani
Tim Butcher	Eni Kume	Sruthy Poulose
Ana Damnjanovic	Jinu Kurian	Emma Thomee
Arvind Arun Dev	Aigars Langins	Florencia Sacarelli
Udara Bimendra	Nikolina Lešić	Abdelqader Zaben

The work of MaMi was wide-ranging and interdisciplinary, covering aspects of magnetism, microfluidics and soft matter. The early-stage researchers enjoyed two extra training periods, one of them at another institute, and the other with an industrial collaborator. Many of chapters in the book are by MaMi participants. Others, on new materials and biological applications, have been added to broaden the scope of the endeavour.

We acknowledge the funding of the MaMi European Union's Horizon 2020 research and innovation programme under grant agreement No. 766007, as well as USIAS and Gutenberg Association of Strasbourg and the Institut Universitaire de France.

Our thanks transcend financial support. The contributions of colleagues from the academic and industrial partners who helped to care for and supervise our students—University of Strasbourg, Institute of Physics and Chemistry of Materials of Strasbourg, Laboratoire Leon Brillouin, Conectus Alsace, microLIQUID S. L., Elvesys SAS, Helmoltz Center Dresden, Max-Planck-Gesellschaft Göttingen, Universidad Del Pais Vasco/ Euskal Herriko Unibertsitatea, Latvijas Universitate, Trinity College Dublin, Qfluidics, Institut Josef Stefan, Kolektor Group—are much appreciated. Special thanks are due to the Centre National de la Recherche Scientifique, for their help in building and organizing the network, as well as the management of the whole process. The work Anna Oleshkevych, Laura Chekli and Peter Dunne in the day-to-day management was invaluable.

Strasbourg, France   Bernard Doudin
Dublin, Ireland   Michael Coey
Riga, Latvia   Andrejs Cēbers

Contents

Contributors

Gwenaël Atcheson AMBER and School of Physics, Trinity College, Dublin, Ireland

G. Bagheri Université de Strasbourg, CNRS UMR 7140, Strasbourg, France

Sylvie Begin Institut de Physique et Chimie des Matériaux, UMR CNRS-UdS 7504, Université de Strasbourg, CNRS, Strasbourg, France

E. Bodenschatz Laboratory for Fluid Physics, Pattern Formation and Biocomplexity, Max Planck Institute for Dynamics and Self-Organization, Göttingen, Germany;
Institute for Dynamics of Complex Systems, University of Göttingen, Göttingen, Germany;
Laboratory of Atomic and Solid State Physics, Cornell University, Ithaca, NY, USA;
Sibley School of Mechanical and Aerospace Engineering, Cornell University, Ithaca, NY, USA

M. Brics MMML Lab, Department of Physics, University of Latvia, Rīga, Latvia

Tim A. Butcher School of Physics and CRANN, Trinity College, Dublin, Ireland; Paul Scherrer Institut, Villigen PSI, Switzerland

Michael Coey AMBER and School of Physics, Trinity College, Dublin, Ireland

Andrejs Cēbers Department of Physics, MMML Lab, University of Latvia, Riga, Latvia

Arvind Arun Dev Université de Strasbourg, CNRS, IPCMS UMR 7504, Strasbourg, France;
School of Applied and Engineering Physics, Cornell University, Ithaca, NY, USA;
Laboratoire Colloïdes et Materiaux Divises, Chemistry Biology & Innovation, ESPCI, PSL Research University, Paris 75005, France

Bernard Doudin IPCMS, University of Strasbourg and Centre National de La Recherche, Strasbourg, France

Peter Dunne Université de Strasbourg, CNRS, IPCMS UMR 7504, Strasbourg, France

Kerstin Eckert Institute of Fluid Dynamics, Helmholtz-Zentrum Dresden-Rossendorf (HZDR), Dresden, Germany;
Institute of Processing Engineering and Environmental Technology, Technische Universität Dresden, Dresden, Germany

Ciaran Fowley Institute of Ion Beam Physics and Materials Research, Helmholtz-Zentrum Dresden-Rossendorf, Dresden, Germany

Barbara Freis NMR and Molecular Imaging Laboratory, Department of General, Organic and Biomedical Chemistry, UMONS, Mons, Belgium;
Institut de Physique et Chimie des Matériaux, UMR CNRS-UdS 7504, Université de Strasbourg, CNRS, Strasbourg, France

Peter A. Galie Department of Biomedical Engineering, Rowan University, Glassboro, NJ, USA

Thomas Gevart NMR and Molecular Imaging Laboratory, Department of General, Organic and Biomedical Chemistry, UMONS, Mons, Belgium

Olav Hellwig Institute of Ion Beam Physics and Materials Research, Helmholtz-Zentrum Dresden-Rossendorf, Dresden, Germany;
Institute of Physics, Chemnitz University of Technology, Chemnitz, Germany

Thomas Hermans Université de Strasbourg, CNRS UMR 7140, Strasbourg, France;
IMDEA Nanociencia, C/ Faraday 9, Madrid, Spain

Paul A. Janmey Departments of Physiology and Physics & Astronomy, University of Pennsylvania, Philadelphia, PA, USA

Aleena Joseph Université de Strasbourg, CNRS, IPCMS UMR 7504, Strasbourg, France

Jinu Kurian CNRS, IPCMS UMR 7504, Université de Strasbourg, Strasbourg, France

Sophie Laurent NMR and Molecular Imaging Laboratory, Department of General, Organic and Biomedical Chemistry, UMONS, Mons, Belgium

Zhe Lei Institute of Fluid Dynamics, Helmholtz-Zentrum Dresden-Rossendorf (HZDR), Dresden, Germany;
Institute of Processing Engineering and Environmental Technology, Technische Universität Dresden, Dresden, Germany

Gerd Mutschke Institute of Fluid Dynamics, Helmholtz-Zentrum Dresden-Rossendorf, Dresden, Germany

Anna Oleshkevych Université de Strasbourg, CNRS, IPCMS UMR 7504, Strasbourg, France

Antonio Ortiz-Ambriz Departament de Física de la Matèria Condensada, Universitat de Barcelona, Barcelona, Spain;
Universitat de Barcelona Institute of Complex Systems (UBICS), Universitat de Barcelona, Barcelona, Spain;
Institut de Nanociència i Nanotecnologia, Universitat de Barcelona, Barcelona, Spain;
Escuela de Ingeniería y Ciencias, Tecnológico de Monterrey, Monterrey, Mexico

Kilian Ortmann Institute of Fluid Dynamics, Helmholtz-Zentrum Dresden-Rossendorf (HZDR), Dresden, Germany

Mattia Ostinato Departament de Física de la Matèria Condensada, Universitat de Barcelona, Barcelona, Spain;
Universitat de Barcelona Institute of Complex Systems (UBICS), Universitat de Barcelona, Barcelona, Spain

O. Petrichenko MMML Lab, Department of Physics, University of Latvia, Rīga, Latvia

Katarzyna Pogoda Department of Experimental Physics of Complex Systems, Institute of Nuclear Physics PAN, Krakow, Poland

Sruthy Poulose School of Physics, Trinity College, Dublin, Ireland

Jennifer A. Quirke School of Physics, Trinity College, Dublin, Ireland

Maria Los Angeles Ramirez Institut de Physique et Chimie des Matériaux, UMR CNRS-UdS 7504, Université de Strasbourg, CNRS, Strasbourg, France

Florencia Sacarelli Université de Strasbourg, CNRS UMR 7140, Strasbourg, France

Ruslan Salikhov Institute of Ion Beam Physics and Materials Research, Helmholtz-Zentrum Dresden-Rossendorf, Dresden, Germany

Dimitri Stanicki NMR and Molecular Imaging Laboratory, Department of General, Organic and Biomedical Chemistry, UMONS, Mons, Belgium

Fengzhi Sun Institute of Fluid Dynamics, Helmholtz-Zentrum Dresden-Rossendorf (HZDR), Dresden, Germany

Rafael Tapia-Rojo Physics Department, King's College London, London, UK

Pietro Tierno Departament de Física de la Matèria Condensada, Universitat de Barcelona, Barcelona, Spain;
Universitat de Barcelona Institute of Complex Systems (UBICS), Universitat de Barcelona, Barcelona, Spain;
Institut de Nanociència i Nanotecnologia, Universitat de Barcelona, Barcelona, Spain

Kiet A. Tran Department of Biomedical Engineering, Rowan University, Glassboro, NJ, USA

Thomas Vangijzegem NMR and Molecular Imaging Laboratory, Department of General, Organic and Biomedical Chemistry, UMONS, Mons, Belgium

Munuswamy Venkatesan AMBER and School of Physics, Trinity College, Dublin, Ireland

Vincent Vivier CNRS, Laboratoire de Réactivité de Surface, UMR 7197, Sorbonne Université, Paris, France

Chapter 1
Introduction

Bernard Doudin, Andrejs Cēbers, and Michael Coey

1.1   Introduction

This introduction situates the subject matter and scope of the book and provides a brief summary of the contents, identifying the magnetic materials and illustrating the range of magnetic effects that can be observed. The importance of size and scaling is emphasised. Some future challenges and prospects are highlighted.

The topic of this collection of articles on applied physics is 'Magnetics and Micro-hydrodynamics', a domain that is situated close to, but separate from three established areas of research (Fig. 1.1)—*Magnetohydrodynamics* [1], the study of the magnetic properties and dynamics of electrically-conducting fluids such as plasmas, liquid metals and ionic solutions, *Ferrohydrodynamics* [2], the study of the motion of strongly polarizable magnetic liquids (ferrofluids) in a magnetic field and *Microfluidics,* the study of fluid flow confined in sub-millimeter scale structures [3]. As investigations accumulate in this generous interstitial space, we see an interesting subfield with characteristics of its own emerging. This book, a collection of original studies and topical reviews is a first attempt to map out the shape of the new subfield. The following 12 chapters are arranged in four groups of three, each part focussed on a different aspect of the physics, chemistry or applications. First we provide thumbnail outlines of the chapters, emphasizing the role of magnetic field in each case.

B. Doudin (✉)
IPCMS, University of Strasbourg and Centre National de La Recherche, Strasbourg, France
e-mail: bernard.doudin@ipcms.unistra.fr

A. Cēbers
Department of Physics, University of Latvia, Riga 1004, Latvia

M. Coey
School of Physics, Trinity College, Dublin 2, Ireland

© The Author(s) 2024
B. Doudin et al. (eds.), *Magnetic Microhydrodynamics*, Topics in Applied Physics 120,
https://doi.org/10.1007/978-3-031-58376-6_1

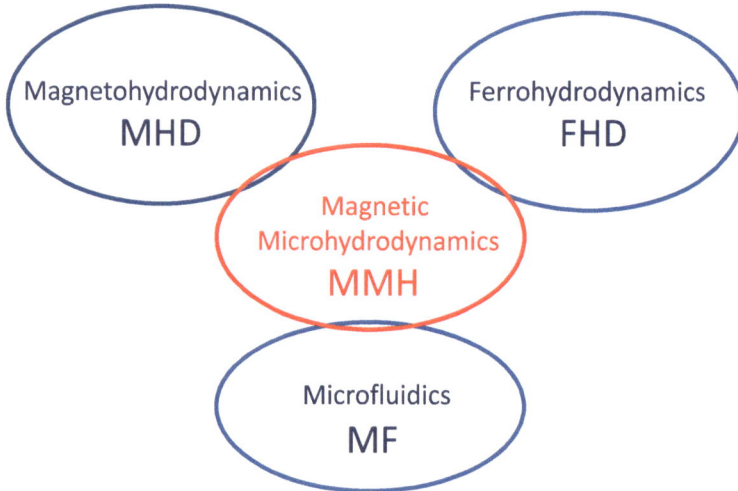

Fig. 1.1 Situation of magnetic microhydrodynamics MMH

The first three articles discuss elements of the theory of *magnetic fields and liquids*, especially the magnetic force and energy densities involved. The first of them, by Tim Butcher on 'Magnetic Action at a Distance', considers the Lorentz force density $f_L = (j \times B)$ where j is the electric current density and the Kelvin or magnetic field gradient force density $f_k = (\chi/2\mu_0)\nabla B^2$ that depends on the susceptibility χ of the medium, provided $\chi \ll 1$ ensuring that the response is linear. (An alternate formulation of the same force is the Korteweg-Helmholtz or concentration gradient force density). The magnetic fields B are ~ 1 T and field gradients range from 1 to 10^7 Tm^{-1}. In an incompressible fluid confined by solid walls, vortex flow arises in systems whenever the gradients of field and susceptibility are noncollinear. A susceptibility of the magnetized medium $\chi \ll 1$ is a common situation. It applies to dilute suspensions of polymeric microbeads loaded with superparamagnetic iron oxide nanoparticles ($\chi \approx 10^{-2}–10^{-3}$), as well as paramagnetic solutions of transition metal ions ($\approx 10^{-3}–10^{-5}$), and water itself ($\chi = -9 \times 10^{-6}$) (Table 1.1).

Gerd Mutschke provides a description of magnetic control of mass transfer in weakly conducting fluids, based on the experience of his group with electrochemical systems, especially for electrolysis of water and metal deposition. The Lorentz force on current flow patterns formed around growing gas bubbles on the electrodes during electrolysis influences hydrogen or oxygen bubble release; oxygen, a weakly paramagnetic gas with $\chi \ll 1$ is subject in addition to a Kelvin force. In metal deposition, local magnetic vortex flow at the electrode stirs and thins the layer of dissolved cations and enhances diffusive mass transport of dissolved metal ions to the cathode where they are reduced to metal. The metal deposits from a mixture of metal ions can be structured by the Kelvin force when the susceptibility and electronegativity of the ions differ.

Table 1.1 Magnetic materials

Material		Size	Susceptibility	Magnetization kAm^{-1}	Coercivity kAm^{-1}
Diamagnets					
Water	Liquid	—	-9×10^{-6}	—	—
Paramagnets					
3d ion solutions	Liquid	—	10^{-5}–10^{-4}	—	—
4f ion solutions	Liquid	—	10^{-4}–10^{-3}	—	—
4f nanoparticles	Solid[b]	20 nm	10^{-3}	—	—
Superparamagnets					
Iron oxide nanoparticles	Solid[b]	5–50 nm	3	400	—
Polymeric microspheres with dispersed iron oxide nanoparticles	Solid[b]	~ 1–3 μm	0.1	30	—
Ferrofluids	Liquid	—	2	30	—
Ferromagnets					
Hematite[a] microparticles Cubes and other shapes	Solid	~ 1–2 μm	10^{-2}	2	300
Iron microspheres Carbonyl iron	Solid[b]	3–10 μm	3	1700	5
Nd-Fe-B magnets Various sizes	Solid	1 mm–5 cm	1	1200	1000

[a] Hematite is a canted antiferromagnet with a very weak ferromagnetic moment
[b] May be dispersed in liquid or soft matter

The third contribution, by Andrejs Cēbers, on phenonomenological models of magnetizable fluids (including ferrofluids with $\chi \approx 1$, beyond the linear approximation) presents several models, considering conservation laws for the energy of the magnetic field and the fluid medium and the mass and momentum of the fluid medium. A choice of the electromagnetic energy flux and stress due to the field yields a relation for entropy production that includes magnetic relaxation. Unlike the other chapters, which use standard SI units, the reader here will encounter the Gaussian cgs units that are still widely used in theoretical magnetism, and will need to replace factors of 4π and c, the velocity of light, by μ_0 and ε_0, the permeability and polarizability of free space [$c^2 = \frac{1}{\mu_0\varepsilon_0}$] to apply them in experimental practice.

Next follows the *Movers and Shakers* section. First is the chapter by Arvind Dev and co-workers, who have looked in detail into the scaling of the liquid-in-liquid flow of water in a moving channel confined by walls of ferrofluid stabilized by a magnetic quadrupole field produced by permanent magnets [4]. Near-ideal plug flow in the 'magnetic antitube' channel is achieved, and the use of surfactant to minimize interfacial tension and reduce the Laplace pressure at the channel wall is explored. Friction is reduced by more than 99% at the moving walls. Shear forces, which are an impediment to moving fragile objects along microchannels, are minimized. X-ray imaging with synchrotron radiation is needed to view the liquid tube behind the ferrofluid walls. The technology can be scaled in principle down to the 1 μm scale.

The first shakers are micron-scale hematite cubes studied by Mārtiņš Brics and co-workers. Hematite is an antiferromagnet where the sublattice moments are slightly canted due to the Dzyaloshinsky-Moriya interaction, creating a weak resultant ferromagnetic moment and a susceptibility two orders of magnitude less than that exhibited by the ferrimagnetic iron oxides magnetite and maghemite. The moment direction lies at $12°$ to a cube diagonal, and it is possible to exert torque on the particles in magnetic field. In a small static field the cubes form straight or kinked chains under the action of dipole–dipole interactions, but in a rotating field the cubes and chains rotate or roll in characteristic ways, including swarming behaviour in two dimensions, that depend on the magnitude and frequency of the rotating magnetic field.

The last article in this section by Mattia Ostinato et al. describes what happens in simulations of two layers of polymeric microparticles loaded with superparamagnetic iron oxide nanoparticles that are confined between glass slides separated by less than two particle diameters. In a low frequency (1 Hz) rotating field, the spheres form dimers that rotate coherently. Then at a critical frequency there is a transition from the coherent state to an 'exchange' state where the dimers begin to break up and reconnect with other partners. The fraction of the particles that are active in this sense is taken as the order parameter of the 'synchronous-exchange' transition, an out-of-equilibrium phase transition belonging to a special universality class that includes other examples such as forest fires and financial crises.

The third section deals with *magnetic field effects on water and ionic solutions*. A wide range of effects on water are reviewed by in the chapter by Sruthy Poulose and co-workers. Effects on the shapes of pendant or sessile droplets are small, because sub-tesla fields have little or no effect on the static or dynamic surface tension of water or ionic solutions. Kelvin forces will modify the droplet shapes, and zero-susceptibility solutions of paramagnetic ions in diamagnetic water are useful to eliminate them. But dramatic field effects of up to 100% or more on the evaporation rate of water in confined spaces cannot be explained by any thermodynamic effect of the field on water chemistry—the magnetic energy is eight orders of magnitude less than the hydrogen bond energies. The explanation is based on field-induced transitions between nuclear *ortho* (triplet) and *para* (singlet) isomers of water with parallel or antiparallel proton spins, which behave as quasi-independent gasses in the vapour phase.

Jinu Kurian et al. use the very large magnetic field gradients created at a multidomain thin film Co–Pt multilayer cathode with perpendicular magnetic anisotropy to influence the model one-electron electrochemical ferricyanide redox reaction at the cathode. The field gradient of order 10^7 Tm^{-1} is localized in the double layer, within about 20 nm of the electrode. The system is investigated by electrochemical impedance spectroscopy. Effects of the magnetic field gradient on the double layer capacitance and the charge transfer resistance are found, but limited to typically 7% when we switch on the Kelvin force by the applied field. Although the field in the double layer is 0.2 T and the average field gradient force is as high as 10^7 Nm^{-3}, they are insufficient to modify greatly diffusion near the electrode. No appreciable influence of the field gradient force on the reaction kinetics was detected.

Fenshe Sun and co-workers describe a new method to measure the rate constant of a rare earth solvent extraction system using an immiscible water–oil system in a thin horizontal Hele-Shaw cell with a 1 mm gap and an interface between Dy(III) in aqueous solution and an organic oil phase. The concentration of dysprosium in the boundary layer is measured interferometrically with a Mach–Zehnder interferometer and the interface mass flux at the initial time is determined for varying Dy(III) and oil concentrations. The system is found to be quasi first-order for Dy and quasi second-order for the oil. The method, which involves no magnetic field, can be used for a wide range of reactions of transparent liquids, and unlike conventional stirring methods, requires only about 1 mL of liquid.

The final section is concerned with *biomedical applications*, which is where the potential impact of bringing magnetism to life on a sub-micron scale is greates. The first two chapters are concerned with the mechanical properties of biomolecules. Peter Galie's chapter on magnetic elastomers and hydrogels for studies of mechanobiology uses synthetic magnetoelastic materials produced by loading PDMS or fibrous polymeric hydrogels with 0.1 wt% of spheres of ferromagnetic carbonyl iron a few microns in diameter. The susceptibility is $\approx 4 \times 10^{-4}$, and the hydrogel is anisotropically stiffened by the magnetic dipole–dipole interaction between the spheres, with a shear modulus of 10^2–10^4 Pa that matches the intercellular medium of soft tissue. Cells are very sensitive to the substrate on which they are grown, and respond in seconds to a change in environmental stiffness by means of intracellular Ca flows. The study compares linear elastomeric and nonlinear fibrous hydrogel systems.

Rafael Tapia-Rojo reports a new design of magnetic tweezers for manipulating biomolecules such as DNA or proteins. Instead of pulling the biomolecule by applying a variable magnetic field gradient from moveable permanent magnets to a micron-scale superparamagnetic polymeric microbead tethered to the molecule, improved control is achieved with a miniature electromagnet in an old-style write head from a tape recorder, which allows fine control of both the magnitude and frequency of the magnetic field gradient force. The method can be used to uncoil proteins and measure their dynamics on a kHz timescale. Two examples are studied. One is protein L where a transient molten globule state appears when the protein is sampled on a millisecond timescale. It is the precursor of the folded states. The other is the Talin mechanosensor, which is exquisitely sensitive to tiny force changes of a few pN, and is shown to exhibit stochastic resonance in the presence of noise.

The last chapter, by Thomas Gevaert and colleagues is an informative review of the dual-purpose magnetic microparticles that can serve both for diagnostics and treatment of disease—an area known as theranostics, which involves pairing of diagnostic biomarkers with therapeutic agents that share a specific target in diseased cells or tissues—cancer in this case. Polymeric microbeads loaded with superparamagnetic iron oxide or gadolinium-based nanoparticles 10–15 nm in diameter can serve as contrast agents in MRI to help locate and identify diseased tissue, and then treat it by heating the local area by means of a kilohertz electromagnetic field in a process known as hypothermia. Drugs attached to the microbead surface may also be released by heating, or used in assay or diagnosis.

Although the applications of magnetics in these chapters are quite diverse, common themes of Lorentz and magnetic field gradient forces recur, both in static and dynamic contexts in low-susceptibility systems. In addition there are dipole–dipole interactions and torques on dilute dispersions of ferromagnetically-ordered inclusions or induced moments in dilute arrays of superparamagnetic nanoparticles in polymeric microbeads. There are all semi-classical effects. To account for effects in magnetochemistry, quantum mechanics and the influence of weak magnetic fields on the symmetry of the electronic or nuclear wave-functions may be critical. Table 1.1 summarizes the properties of various relevant magnetic materials, which are classified by their susceptibility and magnetic order. Size also matters, as the effect of magnetic field and forces will also increase with the size of the object. However, scaling down can be beneficial: the magnetic dipole field is scale independent, hence the field produced by an array of permanent magnets does not depend on the length scale—the secret to the success of magnetic recording. However the gradient field forces are enhanced on decreasing the scale, which allows the creation of intense force fields at the nanoscale, with experimental designs and improved sensitivity illustrated in several chapters of this book.

We hope that these ideas will encourage people with different scientific interests to appreciate the uses and potential of magnetic field and magnetic materials and use them effectively to build a new magnetoscience on sound foundations. A key first step should be to establish experimentally the magnetic principle behind any proposed effect. The experiments can be quite simple at first, but the vector nature of the magnetic field and the combination of magnetic and other forces can complicate the interpretation and modelling of the systems under investigation, with a benefit of a remarkable richness of the observed phenomena. There is therefore a real need to properly identify the forces at play, and search for the best experimental conditions to avoid misinterpretation of the results.

Some areas of study we think will be important for future development of Magnetic Microhydrodynamics (MMH) are the following.

- A focus on the often-elusive influence of magnetic fields and forces on the properties of water, the ubiquitous component of living matter.
- An effort in experimental design that minimizes ambiguity and 'artefacts' in the results.
- Development of models that reveal new behaviour in magnetic microhydrodynamics beyond the linear approximation, analogous to coercivity.
- A truly interdisciplinary approach, where teams are built associating members having an understanding of magnetism with experts in the fields of application such as mechanics, materials chemistry or life science

One example at least serves to emphasis that a fundamental research program in MMH has potential to achieve significant societal relevance: Several authors evoke the idea of efficient magneto-electrochemical separation of the rare-earth elements, the first claims for which date back to the work of Walter and Ida Noddack in the 1950s [5]. It remains a challenge to demonstrate an effective process that can be implemented at scale, which would be a step towards a more sustainable energy landscape where rare-earth permanent magnets will be meeting many of our requirements for magnetic field at no continuous expenditure of energy.

References

1. P.A. Davidson, An Introduction to Magnetohydrodynamics (Cambridge University Press 2010)
2. R.E. Rosensweig, Ferrohydrodynamics (Dover Publications, 2014)
3. P. Tabeling, Introduction to Microfluidics (Oxford University Press, 2010)
4. P. Dunne et al., Nature **581**, 58–62 (2020)
5. W. Noddack, I. Noddack, E. Wicht, Ber. Bunsenges. Phys. Chem.Bunsenges. Phys. Chem. **62**, 77 (1958)

Part I
Magnetism and Liquids

Chapter 2
Magnetic Action at a Distance: Fields, Gradients and Currents in Fluids

Tim A. Butcher

Magnetic fields can alter fluid flow by either interacting with the microscopic magnetic moments of the fluid constituents or with electric currents in a conducting fluid. Control of the flow by magnetic fields is especially convenient, since influencing the flow by mechanical means relies on the interaction of a solid component in direct contact with the fluid. Fluids can be attracted or repelled by a magnetic field gradient depending on whether they are paramagnetic or diamagnetic. The paramagnetic properties depend on the spin configuration of the atoms, ions or molecules that constitute the fluid [1]. A fluid is diamagnetic when its constituents have no unpaired electrons. Magnetic fluids are most commonly encountered in the form of liquids, although paramagnetism also exists in the gaseous state with the prominent example of molecular oxygen [2]. The following sections give a concise overview of the key ideas behind the agglomeration of magnetic particles in magnetic fields, the influence of magnetic field gradients on fluid flows and magnetohydrodynamics.

2.1 Magnetic Liquids

Paramagnetism in the liquid state originates from solvated paramagnetic ions or ferrimagnetic/ferromagnetic particles in colloidal suspensions [3]. In all cases, the solvents or carrier liquids are diamagnetic. Fluids show no net magnetisation without exposure to an external magnetic field. This is due to Brownian motion that keeps the molecules from attaining any ordered state in which the exchange interaction between magnetic particles could have an effect. Brownian motion also ensures the homogeneity of the liquid by keeping the ions or particles from settling. However, the application of a magnetic field gradient to suspensions of magnetic particles

T. A. Butcher (✉)
School of Physics and CRANN, Trinity College, Dublin 2, Ireland
e-mail: tbutcher@tcd.ie

Paul Scherrer Institut, 5232 Villigen PSI, Switzerland

B. Doudin et al. (eds.), *Magnetic Microhydrodynamics*, Topics in Applied Physics 120,
https://doi.org/10.1007/978-3-031-58376-6_2

can lead to their agglomeration in areas of high magnetic field intensity. The magnetically modified distribution $N(H)$ of the density of suspended particles can be estimated from the Boltzmann factor $\exp(-E_{mag}/k_B T)$ with the magnetic energy E_{mag}, Boltzmann constant $k_B = 1.38 \times 10^{-23}\,\mathrm{J\,K^{-1}}$ and temperature T in the exponent. In a suspension made up of particles with volume V and magnetic susceptibility χ, the magnetic energy of the distribution of magnetic moments in an external field is given by $E_{mag} = -\frac{1}{2}\mu_0\chi H^2 V$. The dimensionless volume magnetic susceptibility χ relates the applied magnetic field H (unit: $\mathrm{A\,m^{-1}}$) to the induced magnetisation $M = \chi H$ and $\mu_0 = 4\pi \times 10^{-7}\,\mathrm{N\,A^{-2}}$ is the vacuum permeability. The value of χ determines the form of magnetism of a material. Paramagnetic materials have positive values for the susceptibility $\chi > 0$, whereas diamagnetic substances have negative susceptibilities $\chi < 0$. Thus, the expression for the Boltzmann distribution is simply:

$$N(H) = N(0) \exp\left(\frac{\mu_0\chi H^2 V}{2k_B T}\right), \tag{2.1}$$

with the density of particles in absence of an external magnetic field $N(0)$. Any magnetic interactions between the particles are neglected in this treatment. As soon as the magnetic energy exceeds the thermal energy ($\frac{1}{2}\mu_0\chi H^2 V \geq k_B T$), particles are immobilised and gather in the area of highest magnetic field. Evidently, the agglomeration in a magnetic field is proportional to the volume of the suspended particles and it is possible to estimate the minimum particle size for their magnetic confinement. Rearranging for the diameter d of a spherical particle with volume $V = \frac{4}{3}\pi r^3 = \frac{1}{6}\pi d^3$ leads to:

$$d \geq \left(\frac{12k_B T}{\pi\mu_0\chi H^2}\right)^{\frac{1}{3}}. \tag{2.2}$$

At room temperature, magnetic particles in a suspension with $\chi = 1$ settle in a magnetic field of $0.2\,\mathrm{T}$ ($H = 1.59 \times 10^5\,\mathrm{A\,m^{-1}}$) when their diameter exceeds $8\,\mathrm{nm}$. The expression in Eq. 2.1 was first employed to explain the formation of magnetic powder patterns (Bitter patterns) that form above the domain walls on surfaces of ferromagnetic crystals [4]. The magnetic movement of individual particles lies at the heart of the phenomenon of magnetophoresis, which is particularly relevant for magnetic separation processes.

It is to be noted that the magnetic susceptibility is also temperature dependent itself for paramagnetic particles that follow Curie's law $\chi = \frac{C}{T}$ with the Curie constant C (unit: K). In practice, the amplification of the susceptibility by cooling is limited by the freezing point of the carrier liquid or solvent. Unlike the paramagnetic case, diamagnetic susceptibilities are temperature independent.

The agglomeration of the magnetic constituents of the liquid can be undesired as the heightened density of particles in the magnetic field compresses the fluid and hinders flow. For this reason, ideal ferrofluids comprise magnetic particles with a size below $10\,\mathrm{nm}$ [5] and are superparamagnetic. Solutions of paramagnetic transition metal or rare earth salts have far lower magnetic susceptibilities than ferrofluids and the thermal energy dwarfs the magnetic energy by at least two order of magnitude at room temperature.

Magnetic nanoparticles or solvated paramagnetic ions act like single domain microscopic magnets and the distribution law must be modified. The energy of a magnetic moment in a magnetic field is given by $E = -\mu_0 \mathbf{m} \cdot \mathbf{H} = -\mu_0 m H \cos\theta$. The distribution $N(H)$ of single domain magnetic nanoparticles or ions in a magnetic field can be obtained by integrating the Boltzmann factor $\exp(-E_{\text{mag}}/E_{\text{therm}}) = \exp(\mu_0 m H \cos\theta / k_B T)$ over all angles between $0°$ and $180°$ [4]:

$$N(H) = N(0) \int_0^{\pi} \sin(\theta) \exp(\mu_0 m H \cos\theta / k_B T) d\theta \tag{2.3}$$

$$= N(0) \left[-\frac{\exp(\frac{\mu_0 m H \cos\theta}{k_B T})}{\mu_0 m H / k_B T} \right]_0^{\pi} \tag{2.4}$$

$$= N(0) \frac{\sinh(\mu_0 m H / k_B T)}{\mu_0 m H / k_B T}. \tag{2.5}$$

Once again, the thermal energy must be larger than the magnetic energy to guarantee homogeneity in the magnetic field. A maximum value of the magnetic field can be estimated from $k_B T \geq \mu_0 m H$. For a gadolinium ion with $7\,\mu_B$ (Bohr magneton: $\mu_B = 9.274 \times 10^{-24} \text{J T}^{-1}$), this yields $\mu_0 H = \frac{k_B T}{m} = \frac{4.11^{-21}}{7 \times 9.274 \times 10^{-24}} \text{T} \approx 63.4\text{T}$. Continuous magnetic fields of such high intensities are not attainable, but pulsed field magnets can sustain such high magnetic fields for milliseconds. In ideal ferrofluids and paramagnetic salt solutions there is no danger of segregation of the magnetic nanoparticles or ions. However, the microscopic forces on the individual microscopic moments are passed on to the bulk fluid as a whole and the resulting force density can drive convection. An explanation of the effects of such external body forces on fluids can be found in fluid dynamics.

2.2 Fluid Dynamics

The Navier-Stokes equation describes the movement of a fluid that is exposed to force densities $\mathbf{F}$ (in N m^{-3}) and pressure gradients ∇P. It is also known as the momentum equation. The change of the fluid velocity $\mathbf{u}$ is introduced into the equation via its material derivative ($\frac{D\mathbf{u}}{Dt} = \frac{\partial \mathbf{u}}{\partial t} + (\mathbf{u} \cdot \nabla)\mathbf{u}$):

$$\rho \frac{D\mathbf{u}}{Dt} = -\nabla P + \eta \nabla^2 \mathbf{u} + \mathbf{F}, \tag{2.6}$$

with the density ρ (in kg m^{-3}) and dynamic viscosity η (unit: N s m^{-2}). The viscosity can be interpreted to cause momentum to diffuse along its gradient [6].

A dimensionless quantity central to fluid dynamics is the Reynolds number (Re), which is the ratio between inertia and viscous forces that appear in the Navier-Stokes equation (Eq. 2.6) as $\rho(\mathbf{u} \cdot \nabla)\mathbf{u}$ and $\eta\nabla^2\mathbf{u}$, respectively. The Reynolds number dictates whether fluid flow is laminar or turbulent and is defined as:

$$\mathrm{Re} = \frac{\rho u l}{\eta} = \frac{u l}{\nu}, \tag{2.7}$$

with the characteristic length scale of the fluid motion l, the fluid velocity u and the kinematic viscosity ν (in m^2 s^{-1}). Low Reynolds numbers indicate laminar flow that is dominated by viscous forces. Laminar flow is widespread in microfluidics due to the small diameter of the tubes in microfluidic circuits. Turbulent flow arises at high Reynolds numbers, at which inertial forces prevail.

Relevant for the following discussion of magnetic effects are the body forces, which enter into the equation as force densities. The most common force density on a fluid is due to gravity:

$$\mathbf{F}_\mathrm{g} = \rho\mathbf{g}, \tag{2.8}$$

where $\mathbf{g}$ is the gravitational acceleration ($g = 9.81$ m s^{-2}). The magnetic force density exerted on a fluid by a magnetic field gradient $\nabla\mathbf{H}$ (in A m^{-2}) is given by the Kelvin force:

$$\mathbf{F}_\mathrm{K} = \mu_0(\mathbf{M} \cdot \nabla)\mathbf{H}. \tag{2.9}$$

For low susceptibility liquids ($\chi \ll 1$) such as paramagnetic salt solutions, an approximated form of the Kelvin force with the magnetic flux density $\mathbf{B} = \mu_0(\mathbf{H} + \mathbf{M})$ (unit: T) is often encountered:

$$\mathbf{F}_{\nabla\mathrm{B}} = \frac{\chi}{\mu_0}(\mathbf{B} \cdot \nabla)\mathbf{B}. \tag{2.10}$$

The approximation $\mathbf{B} = \mu_0(\mathbf{H} + \mathbf{M}) = \mu_0(\mathbf{H} + \chi\mathbf{H}) \approx \mu_0\mathbf{H}$ is used to obtain this expression, which is not applicable to ferrofluids with $\chi \approx 1$.

2.2.1 Convection in a Magnetic Field Gradient

What effect do these force densities have upon fluids? Magnetic field gradients can deform free surfaces of liquids, but the discussion will be restricted to cases in which the fluid is completely confined by solid walls, as is the case in microfluidic systems. Furthermore, the fluid is assumed to be an incompressible liquid. This is a reasonable assumption for ideal ferrofluids and paramagnetic salt solutions. For incompressible fluids, the flow velocity is free of divergence:

$$\nabla \cdot \mathbf{u} = 0. \tag{2.11}$$

Whether a force density has any influence on the flow velocity of an enclosed liquid depends on the vorticity $\boldsymbol{\omega}$ (unit: s^{-1}), which quantifies the spinning motion of a fluid. The vorticity is defined as the curl of the flow velocity:

$$\boldsymbol{\omega} = \nabla \times \mathbf{u}. \tag{2.12}$$

The curl in Eq. 2.12 evinces the three-dimensionality of the vorticity. Two-dimensional approximations can lead to misleading results and should generally be avoided. If a force density does not induce vorticity in an enclosed incompressible fluid, there is no noticeable effect. The force only presses the fluid against the solid wall, resulting in a pressure field that cancels it out. The density of an incompressible fluid is unperturbed by this and nothing happens. No irrotational flow is created. In contrast, internal flows can materialise in the fluid when vorticity is present. This is known as rotational flow. The Navier-Stokes equation can be transformed into the vorticity equation by application of the curl operator and division by ρ [7, 8]:

$$\frac{D\boldsymbol{\omega}}{Dt} = (\boldsymbol{\omega} \cdot \nabla)\mathbf{u} + \nu\nabla^2\boldsymbol{\omega} + \frac{1}{\rho^2}\nabla\rho \times \nabla p + \nabla \times \left(\frac{\mathbf{F}}{\rho}\right), \tag{2.13}$$

with the material derivative of the vorticity ($\frac{D\boldsymbol{\omega}}{Dt} = \frac{\partial \mathbf{u}}{\partial t} + \mathbf{u} \cdot \nabla\boldsymbol{\omega}$). Here, the kinematic viscosity $\nu = \frac{\eta}{\rho}$ (in $m^2 s^{-1}$) has also been introduced. For compressible fluids, the term $\boldsymbol{\omega}(\nabla \cdot \mathbf{u})$ must be added to the equation. The first term on the right ($\boldsymbol{\omega} \cdot \nabla)\mathbf{u}$ describes how gradients of the flow velocity influence the vorticity, whereas the second term ($\nu\nabla^2\boldsymbol{\omega}$) acts as the diffusion of vorticity down its gradient due to viscosity. The third term is relevant for stratified fluids with $\nabla\rho \neq 0$ in which the pressure field interacts with the density gradient. The curl of the force densities is the last term and this is where magnetic fields come into play. Hence, it is sufficient to analyse the curl of the body forces for incompressible fluids that are enclosed by solid walls.

The curl of the gravitational force density (Eq. 2.8) is given by:

$$\nabla \times \mathbf{F}_g = \nabla\rho \times \mathbf{g}. \tag{2.14}$$

It follows that the gravitational force density is irrotational when the density gradient is parallel to the direction of gravity. Gravity induces flows in any system in which this is not the case. The effect of a magnetic field gradient can be analysed by means of the rotational component of the Kelvin force [3, 9]:

$$\nabla \times \mathbf{F}_K = \mu_0 \nabla M \times \nabla H. \tag{2.15}$$

This implies that there must be an *inhomogeneity in the magnetisation, as well as in the applied magnetic field.* Additionally, the gradients of both must be non-parallel. As the magnetisation of paramagnetic liquids is proportional to the concentration of their magnetic component, a gradient in the density usually accompanies the magnetisation gradient. Thus, an interplay of gravitational (Eq. 2.14) and magnetic force densities (Eq. 2.15) often develops in systems with an inhomogeneity. Large gradients in the magnetisation are located at the interface between magnetic and non-magnetic liquids. In the case of miscible liquids, such an interface is inevitably wiped out by molecular diffusion that establishes homogeneity and brings the system into thermodynamic equilibrium [10, 11]. On the other hand, the magnetic component in a system of immiscible liquids can be captured in a magnetic field gradient indefinitely [12].

The role of the concentration c (in $mol\,m^{-3}$) dependence of the magnetisation is easy to identify by inspecting the curl of the magnetic field gradient force (Eq. 2.10) when it is formulated with the molar magnetic susceptibility χ_m (in $m^3\,mol^{-1}$):

$$\mathbf{F}_{\nabla B} = \frac{\chi_m\,c}{\mu_0}(\mathbf{B}\cdot\nabla)\mathbf{B} = \frac{\chi_m\,c}{2\mu_0}\nabla B^2. \tag{2.16}$$

The identity $(\mathbf{B}\cdot\nabla)\mathbf{B} = \frac{1}{2}\nabla(\mathbf{B}\cdot\mathbf{B}) - \mathbf{B}\times(\nabla\times\mathbf{B})$ with the approximation $\nabla\times\mathbf{B} = 0$ results in the right part of Eq. 2.16. This is valid for $\mathbf{B}\approx\mu_0\mathbf{H}$, when currents are absent and $\nabla\times\mathbf{H} = 0$. Applying the curl operator to Eq. 2.16 and using the identity $\nabla\times(\psi\nabla\phi) = \nabla\psi\times\nabla\phi$ leads to:

$$\nabla\times\mathbf{F}_{\nabla B} = \frac{\chi_m}{2\mu_0}\nabla c\times\nabla B^2. \tag{2.17}$$

At this point it is worth mentioning that there is another prevalent expression for the magnetic force density known as the Korteweg-Helmholtz force [3, 13]. It is defined in terms of gradients of the magnetic susceptibility or permeability:

$$\mathbf{F}_H = -\frac{\mathbf{H}\cdot\mathbf{H}}{2}\mu_0\nabla\chi = -\frac{1}{2}H^2\mu_0\nabla\chi. \tag{2.18}$$

The approximation of the Korteweg-Helmholtz force (Eq. 2.18) for fluids with modest magnetic susceptibilities $\chi\ll 1$ is called the concentration gradient force:

$$\mathbf{F}_c = -\frac{1}{2}H^2\mu_0\chi_m\nabla c \approx -\frac{B^2}{2\mu_0}\chi_m\nabla c. \tag{2.19}$$

The expressions for the Kelvin (Eq. 2.9) and the Korteweg-Helmholtz force (Eq. 2.18) differ by a gradient of the pressure and offer identical alternative descriptions for the motion of incompressible fluids in magnetic field gradients [3, 14, 15]. Both sets of equations describe the creation of vorticity in a magnetic field gradient in the same way. This is straightforward to verify by assuring oneself that the application of the curl operator to the Korteweg-Helmholtz force (Eq. 2.18) leads to the rotational component of the Kelvin force (Eq. 2.15). Likewise the curl of the concentration gradient force (Eq. 2.19) corresponds to that of the magnetic field gradient force (Eq. 2.17). The vector identity $\nabla \times (\psi \nabla \phi) = \nabla \psi \times \nabla \phi$ is needed to prove this.

If a fluid exhibits a free surface and is exposed to a magnetic field, the pressure at the interface can be modified by the magnetic force density and deform the surface. This is pertinent for the magnetic control of the wetting of a solid or the deformation of a magnetisable liquid drop [16]. Here, irrotational force densities contribute and potential flow ensues. The situation for incompressible fluids and irrotational flows is described by the time-dependent ferrohydrodynamic Bernoulli equation [5, 16]. However, rotational flows still dominate the flow field in an open beaker of magnetic fluid when the magnetic field is applied far from the surface of the liquid towards the bottom of the vessel. Such conditions are typical in magnetoelectrochemistry, where the electrodes are completely submerged in the electrolyte.

2.2.2 Magnetohydrodynamics

Magnetic fields can also interact with electrical currents in fluids and modify fluid flow through the Lorentz force. Magnetohydrodynamics (MHD) is the field of study that deals with the effect of the Lorentz force on the motion of electrically conducting fluids [17]. Electrodes can pass a current through electrolytic solutions, which can then interact with an externally applied magnetic field and drive flow. The Lorentz force density is defined as the cross product of the current density $\mathbf{j}$ (in $\mathrm{A\,m^{-2}}$) and the magnetic flux density $\mathbf{B}$:

$$\mathbf{F}_L = \mathbf{j} \times \mathbf{B}. \tag{2.20}$$

The current and magnetic flux densities must be non-parallel for the generation of a Lorentz force component. An important dimensionless number quantity in MHD is the Hartmann number (Ha), which is the square root of the ratio between the Lorentz force and the viscous forces [17]:

$$\mathrm{Ha} = Bl\sqrt{\frac{\sigma}{\eta}}, \tag{2.21}$$

with the electrical conductivity σ (in $\Omega^{-1}\,\mathrm{m^{-1}}$), the magnetic flux density B, the dynamic viscosity η and the characteristic length scale l as in the definition of the Reynolds number (Eq. 2.7). The Hartmann number is proportional to the magnitude of the magnetic flux density B, but is diminished at small length scales. This means that the Lorentz force becomes negligible at the microscale in all but the highest conductivity liquids, regardless if the Lorentz force is curl-free or not. It is worth pointing out that the Kelvin force (Eq. 2.9) is more resilient to miniaturisation, as the smaller length scales increase the magnetic field gradients. A comparison between the magnitudes of the Kelvin and Lorentz force densities is given in Table 2.1 at the end of this chapter.

Electrolytic cells or vessels containing liquid metals are bounded by solid walls and the rotational component of the Lorentz force is necessary to introduce vorticity into the electrolyte solution. Utilising the vector identity for the curl of a cross product $\nabla \times (\mathbf{A} \times \mathbf{B}) = (\mathbf{B} \cdot \nabla)\mathbf{A} - (\mathbf{A} \cdot \nabla)\mathbf{B} + \mathbf{A}(\nabla \cdot \mathbf{B}) - \mathbf{B}(\nabla \cdot \mathbf{A})$ leads to:

$$\nabla \times \mathbf{F}_L = \nabla \times (\mathbf{j} \times \mathbf{B}) = (\mathbf{B} \cdot \nabla)\mathbf{j} - (\mathbf{j} \cdot \nabla)\mathbf{B} + \mathbf{j}(\nabla \cdot \mathbf{B}) - \mathbf{B}(\nabla \cdot \mathbf{j}). \qquad (2.22)$$

The last two terms drop out because of Gauss's law for magnetism $\nabla \cdot \mathbf{B} = 0$ and charge conservation $\nabla \cdot \mathbf{j} = 0$, respectively. This leaves only the first two terms:

$$\nabla \times \mathbf{F}_L = (\mathbf{B} \cdot \nabla)\mathbf{j} - (\mathbf{j} \cdot \nabla)\mathbf{B}. \qquad (2.23)$$

Accordingly, there must be an inhomogeneity in the electrical current or the external magnetic field in order for a rotational component of the Lorentz force to exist. This is different from the rotational component of the Kelvin force (Eq. 2.15), which also relies on a gradient in the magnetisation provided by concentration gradients (Eq. 2.17). A three-dimensional analysis of the situation in electrolytic cells is mandatory for reliable interpretations of experimental results [18]. When an electrical current flows through a paramagnetic electrolytic solution in a magnetic field gradient, both the Kelvin force and the Lorentz force are present. Which of these is more relevant for the motion of the electrolyte depends on the magnetic susceptibility and electrical conductivity of the liquid.

It is also possible to influence fluid flow with the Lorentz force and no input of electrical energy. A fluid of high electrical conductivity such as liquid metal can be moved through a magnetic field, which induces an electrical current proportional to $\sigma(\mathbf{u} \times \mathbf{B})$ and the flow velocity $\mathbf{u}$. Liquid metals have a conductivity of around $10^6 \, \Omega^{-1} \, \mathrm{m}^{-1}$, while that of sea water is approximately $5 \, \Omega^{-1} \, \mathrm{m}^{-1}$ for comparison. Induced currents in liquid metals moving at $1 \, \mathrm{m \, s}^{-1}$ through an external field of $1 \, \mathrm{T}$ magnitude are around $10 \, \mathrm{A \, cm}^{-2}$. The appearance of these currents and the subsequent dissipation of kinetic energy finds application in metallurgical processes, where motion of the melt can be controlled by magnetic damping [17]. The magnitude and distribution of the magnetic damping force can be estimated with $\sigma(\mathbf{u} \times \mathbf{B}) \times \mathbf{B}$. In the case of a flowing electrolytic solution in an external magnetic field, the induced currents are negligible.

Another dimensionless number carrying the name of interaction parameter or Stuart number (N) serves to gauge if the Lorentz force due to currents induced by motion of a liquid in an external magnetic field have an effect upon the flow. The interaction parameter is defined as the ratio between the Lorentz force $\mathbf{j} \times \mathbf{B}$, with $|\mathbf{j}| \sim \sigma u B$, and inertia $\rho(\mathbf{u} \cdot \nabla)\mathbf{u}$:

$$N = \frac{\sigma B^2 l}{\rho u} = \frac{\mathrm{Ha}^2}{\mathrm{Re}}. \qquad (2.24)$$

Table 2.1 Magnitudes of force densities acting on paramagnetic salt solutions ($\chi = 10^{-3}$), diamagnetic liquids ($\chi = -10^{-5}$) and ferrofluids ($\chi = 1$) with following parameters: $B = 1\,\text{T}$, $\nabla B = 10\,\text{T}\,\text{m}^{-1}$, $j = 10^3\,\text{A}\,\text{m}^{-2}$ and $\Delta\rho = 10^2\,\text{kg}\,\text{m}^{-3}$

Force density	Equation	Fluid	Magnitude ($\text{N}\,\text{m}^{-3}$)
Kelvin	$\frac{\chi}{\mu_0}B\nabla B$	Paramagnetic solution	10^4
		Diamagnetic liquid	10^2
		Ferrofluid	10^6
Lorentz	$\mathbf{j} \times \mathbf{B}$	Salt solution	10^3
Gravitational	$\Delta\rho\mathbf{g}$	Salt solution	10^3
Magnetic damping	$\sigma(\mathbf{u} \times \mathbf{B}) \times \mathbf{B}$	Salt solution	10^1
		Liquid metal	10^5

The magnetic damping force in a liquid metal ($\sigma = 10^6\,\Omega\,\text{m}^{-1}$) and salt solution ($\sigma = 10^2\,\Omega\,\text{m}^{-1}$) flowing through a 1 T magnetic field at $u = 10^{-1}\,\text{m}\,\text{s}^{-1}$ is also included. Although the magnitudes of the forces indicate the strength of their overall effect, it is the force distributions that govern the deformation of the fluid

The interaction parameter equals the ratio of the squared Hartmann number (Eq. 2.21) and the Reynolds number (Eq. 2.7). All three dimensionless quantities are reduced when the characteristic length l reaches small values. Unlike the Hartmann number, the interaction parameter is proportional to the square of the magnetic flux density. The ratio given by Eq. 2.24 indicates that viscous forces dominate the Lorentz force at high flow velocities, even if the induced current is directly proportional to u.

It is possible to express the Lorentz force (Eq. 2.20) solely with the magnetic flux density $\mathbf{B}$ using Ampère's law[1] $\nabla \times \mathbf{B} = \mu_0\mathbf{j}$ [17, 19]:

$$\mathbf{F}_{\text{L}} = \mathbf{j} \times \mathbf{B} = \frac{1}{\mu_0}(\mathbf{B} \cdot \nabla)\mathbf{B} - \frac{\nabla B^2}{2\mu_0}. \tag{2.25}$$

The first term is called the magnetic tension and the second is the gradient of the magnetic pressure. The former can have a non-zero rotational component $\nabla \times (\frac{1}{\mu_0}(\mathbf{B} \cdot \nabla)\mathbf{B})$, whereas the latter has none. In any case, the existence of a gradient of the magnetic flux density $\nabla B \neq 0$ is a prerequisite for the generation of vorticity in the liquid. On a final note, it is interesting that the magnetic tension and the approximation that is the magnetic field gradient force (Eq. 2.10) differ only by the factor of the magnetic susceptibility.

---

[1] In matter, the magnetic flux density is given by $\mathbf{B} = \mu_0(\mathbf{H} + \mathbf{M})$ and the current can be decomposed into $\mathbf{j} = \mathbf{j}_{\text{free}} + \mathbf{j}_{\text{bound}}$. Ampère's law $\nabla \times \mathbf{B} = \mu_0\mathbf{j}$ shows that $\mathbf{H}$ is associated with the measurable free currents $\nabla \times \mathbf{H} = \mathbf{j}_{\text{free}}$ and $\mathbf{M}$ with the non-dissipative magnetisation currents $\nabla \times \mathbf{M} = \mathbf{j}_{\text{bound}}$.

2.3 Summary

The present chapter gave an overview of the effect of magnetic fields on magnetic fluids and those with a non-negligible conductivity. Gradients in the magnetic field can affect fluid flow and even manipulate magnetic particles suspended in liquids. The magnetic particles must have sufficient size in order for a magnetic field to change their concentration. This is important for the phenomena of magnetorheology [20, 21], in which the viscosity of the fluid is magnetically modified, and magnetophoresis [21, 22], where particles migrate in a magnetic field. Magnetic fluids with particles permanently in suspension without sedimentation are incompressible. When an incompressible fluid is confined by solid walls, the rotational component of the Kelvin force can drive convection. For this to happen, there must be a gradient both in the magnetic field and the magnetisation. When conducting fluids move through a magnetic field, the Lorentz force can alter the fluid flow and lead to stirring or magnetic damping. Magnitudes of the different force densities present in fluids are summarised in Table 2.1. Whether or not the magnetic forces still have an effect upon scaling down to a microfluidic system depends on their strength with respect to the friction given by the viscous forces. In any case, maximisation of the magnetic susceptibility of the liquid, design of high magnetic field gradients and use of a highly conducting liquid for MHD flow will serve to exploit the magnetic field to greater effect.

Acknowledgements Support from the European Commission under contract No. 766007 for the MAMI Marie Curie International Training Network and the Swiss Nanoscience Institute (SNI) is acknowledged.

References

1. J.M.D. Coey, *Magnetism and Magnetic Materials*, chapter 4, 1st edn. (Cambridge University Press, 2010)
2. P.W. Selwood, *Magnetochemistry* (Interscience Publishers Inc., New York, 1943)
3. T.A. Butcher, J.M.D. Coey, Magnetic forces in paramagnetic fluids. J. Phys.: Condens. Matter. **35**(5), 053002 (2023). https://doi.org/10.1088/1361-648X/aca37f
4. C. Kittel, Theory of the formation of powder patterns on ferromagnetic crystals. Phys. Rev. **76**, 1527 (1949). https://doi.org/10.1103/PhysRev.76.1527
5. R.E. Rosensweig, *Ferrohydrodynamics* (Dover Publications, Incorporated, 1998)
6. D.J. Tritton, *Physical Fluid Dynamics*, 2nd edn (Oxford University Press, 1988)
7. D.J. Tritton, *Physical Fluid Dynamics*, chapter 6, 2nd edn (Oxford University Press, 1988), pp. 81–88
8. C. Truesdell, *The Kinematics of Vorticity* (Science series. Indiana University Press, Indiana University publications, 1954), 9780598613103
9. E. Blums, *Andrejs Cēbers, and Mikael M Maiorov* (Walter de Gruyter, Magnetic fluids, 1996)
10. J.M.D. Coey, R. Aogaki, F. Byrne, P. Stamenov, Magnetic stabilization and vorticity in submillimeter paramagnetic liquid tubes. Proc. Natl. Acad. Sci. U.S.A. **106**(22), 8811–8817 (2009). ISSN 0027-8424. https://doi.org/10.1073/pnas.0900561106, https://www.pnas.org/content/106/22/8811

11. T.A. Butcher, G.J.M. Formon, P. Dunne, T.M. Hermans, F. Ott, L. Noirez, J.M.D. Coey, Neutron imaging of liquid-liquid systems containing paramagnetic salt solutions. Appl. Phys. Lett. **116**(2), 022405 (2020). https://doi.org/10.1063/1.5135390

12. P. Dunne, T. Adachi, A.A. Dev, A. Sorrenti, L. Giacchetti, A. Bonnin, C. Bourdon, P.H. Mangin, J.M.D. Coey, B. Doudin, et al., Liquid flow and control without solid walls. Nature **581**(7806), 58–62 (2020). https://doi.org/10.1038/s41586-020-2254-4

13. J.R. Melcher, *Continuum Electromechanics*, chapter 3 (MIT Press Cambridge, 1981)

14. N. Bobbio, S. Bobbio, *Electrodynamics of Materials: Forces, Stresses, and Energies in Solids and Fluids* (Taylor & Francis US, 2000)

15. P. Mazur, S.R. de Groot, On pressure and ponderomotive force in a dielectric statistical mechanics of matter in an electromagnetic field ii. Physica **22**(6), 657–669 (1956). ISSN 0031-8914, https://doi.org/10.1016/S0031-8914(56)90014-3, https://www.sciencedirect.com/science/article/pii/S0031891456900143

16. Y. Zimmels, The Bernoulli equation for fluids in electromagnetic and interfacial systems. J. Colloid Interface Sci. **125**(2), 399–419 (1988). https://doi.org/10.1016/0021-9797(88)90004-5

17. P.A. Davidson, An introduction to magnetohydrodynamics, in *Cambridge Texts in Applied Mathematics* (Cambridge University Press, 2001). https://doi.org/10.1017/CBO9780511626333

18. G. Mutschke, A. Bund, On the 3D character of the magnetohydrodynamic effect during metal electrodeposition in cuboid cells. *Electrochem. Commun.* **10**(4), 597–601 (2008). ISSN 1388-2481, https://doi.org/10.1016/j.elecom.2008.01.035, https://www.sciencedirect.com/science/article/pii/S1388248108000374

19. J.A. Shercliff, The dynamics of conducting fluids under rotational magnetic forces. Sci. Prog. **66**(262), 151–170 (1979). ISSN 00368504, 20477163, http://www.jstor.org/stable/43420487

20. J.R. Morillas, J. de Vicente, Magnetorheology: a review. Soft Matter **16**, 9614–9642 (2020), https://doi.org/10.1039/D0SM01082K

21. N.-T. Nguyen, Micro-magnetofluidics: interactions between magnetism and fluid flow on the microscale. Microfluid. Nanofluidics **12**(1–4), 1–16 (2012). https://doi.org/10.1007/s10404-011-0903-5

22. F. Alnaimat, S. Dagher, B. Mathew, A. Hilal-Alnqbi, S. Khashan, Microfluidics based magnetophoresis: a review. Chem. Rec. **18**(11), 1596–1612 (2018). https://doi.org/10.1002/tcr.201800018

Chapter 3
Magnetic Control of Flow and Mass Transfer in Weakly Conducting Fluids

Gerd Mutschke

3.1 Introduction

This review summarizes work carried out over the past 20 years to utilize magnetic fields for controlling flow and mass transfer in weakly conducting fluids. It will mainly focus on lab-scale (dimension $L < 1$ m) applications in aqueous solutions, e.g. electrolytes or sea water, with a typical electrical conductivity σ of about 1–10 S/m. Control can often be achieved by utilizing the Lorentz force density, which is defined as the vector product of current density j in the liquid and the magnetic induction B [1]. As typical flow velocities and magnetic fields will be $u < 1$ m/s, B < 1 T, the current density induced by flow $\sigma(u \times B)$ might be too small to achieve control. Therefore, in the following we consider cases where an external current density j_0 is applied, compared to which the induced current density can be neglected. Furthermore, the magnetic Reynolds number $R_m = \mu_0 \sigma u L$ will be small (μ_0 is the vacuum permeability), and the induced magnetic field can be neglected compared to the applied field B_0 [1]. The Lorentz force density therefore reads

$$f_L = j_0 \times B_0 \tag{3.1}$$

Utilizing the Lorentz force mostly relies on adding momentum to the electrolyte to energize wall-parallel flow, which then also impacts mass transfer at electrodes. The second magnetic force considered is the magnetic gradient force or Kelvin force density, which requires the fluid to be paramagnetic or diamagnetic and a spatial gradient of the magnetic field applied. It reads [2]

Revised version, 22.05.2023

G. Mutschke (✉)
Institute of Fluid Dynamics, Helmholtz-Zentrum Dresden-Rossendorf, Bautzner Landstr. 400, 01328 Dresden, Germany
e-mail: g.mutschke@hzdr.de

© The Author(s) 2024
B. Doudin et al. (eds.), *Magnetic Microhydrodynamics*, Topics in Applied Physics 120,
https://doi.org/10.1007/978-3-031-58376-6_3

$$f_K = \frac{\chi_{sol}}{\mu_0} B_0 \nabla B_0 \tag{3.2}$$

The magnetic susceptibility χ_{sol} of the solution comprises the diamagnetic contribution of the water molecules and the concentration-dependent contribution of further particles or ions. It reads

$$\chi_{sol} = \chi_{H2O} + \sum_i \chi_i^{mol} c_i \tag{3.3}$$

where c_i denotes the molar concentration of species i in the solution. Parts of the force named concentration gradient force in earlier literature can often be neglected, as clarified in [3]. Although it is not discussed here, it should also be mentioned that effects on particles in liquids (e.g. magnetic torque) can be achieved by magnetic fields, and that the dipole–dipole interaction force can be utilized for magnetic control. For further details and examples see [2, 4]. As chemical reactions taking place at electrodes or an existing spatial variation of the electric current density in the electrolyte may change the density of the electrolyte, effects of solutal and thermal buoyancy may need to be considered as well.

In the following, applications in flow control and electrochemical processes will be reviewed. The latter are of specific interest, as the electric current necessary for Lorentz force control is an intrinsic part of the system and does not to have to be applied externally. What will be considered is water electrolysis to create shear flows by the Lorentz force to enhance bubble departure from the electrode and electrochemical metal deposition where the space–time yield and the uniformity of the deposit can be enhanced. Furthermore, structuring of electrodeposits down to the nanoscale can be improved by applying magnetic fields. For the sake of brevity, the references given mostly focus on own work, in which a broader view on the topics may also be found. Finally, it should be mentioned that apart from the applications given below, the magnetic gradient force can further be used to control interfaces of paramagnetic liquids [5, 6], to enhance the volumetric enrichment of rare earth elements in solution [7, 8] or to manipulate flow driven in microfluidic channels for analysis purposes [9].

3.2 Flow Control

First attempts to utilize the Lorentz force for propelling ships date back to the 1960s in the US for experimental submarines. The basic principle is that electrical current from wall-mounted electrodes at opposite sides and a perpendicularly oriented magnetic field generate thrust in a channel [10]. Apart from problems related to gas evolution and corrosion issues, the efficiency of those MHD thrusters is typically low in electrolytes, as the power loss per volume scales as j^2/σ. Therefore, applying strong magnetic fields is of advantage. In 1994, the MHD ship "YAMATO-1" tested in Japan was the first to use superconducting magnets for propulsion [11]. Recent developments in designs for propelling cruise ships are reported in [12]. As the total losses remain considerable, a way out is to reduce the actuation volume. In that sense, boundary layer control, well studied in aerodynamics for flaps and airfoils, by electromagnetic means appears to be more attractive. Based on a setup of surface flush-mounted alternating stripes of electrodes and magnets [13] the boundary layer can be energized by adding streamwise momentum, whereby the depth of forcing can be easily controlled by the width of the stripes. The Lorentz force created by the actuator exponentially decays in wall-normal direction, which was found to be advantageous for stabilizing the boundary layer and delaying transition to turbulence [14, 15]. Several efforts have been undertaken to control the separation of flow at inclined hydrofoils, useful for stabilizing large cruise ships. The area of control can thereby be restricted to a small front part at the suction side of the foil. In order to further improve the efficiency of control, proper time-oscillating instead of static forcing can be applied to reduce the separation and thus to enhance the rudder efficiency [16, 17]. Although the method works well also in turbulent flow at high Reynolds numbers, its application remains limited due to the strong increase of energy consumption with rising flow velocity [18].

3.3 Gas Evolution During Water Electrolysis

Electrochemical splitting of water into hydrogen and oxygen today appears attractive for storing the fluctuating renewable energy supply in chemical form. The efficiency of alkaline water electrolysis at high current densities suffers from intense gas evolution, thereby increasing ohmic and kinetic losses [19]. At vertical electrodes, beside buoyancy and pumping the electrolyte, magnetic fields allow generation of a Lorentz force that drives additional shear to enhance upward bubble transport [20]. Apart from the behavior at large planar electrodes, the phenomena at microelectrodes are also of interest as they could serve as a generic model for electrodes with small islands of catalytic materials or surface elevations at the micro- or nano-scale. Here it was found that surface-parallel magnetic fields generate a shear flow similar to the behavior at large electrodes that strongly favours small-size bubble departure [21]. Surface-normal magnetic fields can be expected to generate counter-rotating azimuthal flows of different strength around the bubble in the upper and lower hemisphere due to tangential deflection of the current density vectors near the bubble surface. At large electrodes, a faster bubble departure is observed, whereas at microelectrodes their departure is retarded. Here, the azimuthal flow driven in the lower hemisphere was found to dominate. Then, due to centrifugal acceleration, a secondary flow is driven towards the bubble/electrode, which retards the bubble departure [22, 23]. Finally, it should be mentioned that inhomogeneous magnetic fields may also appear from magnetization of the electrodes and influence the evolution of paramagnetic oxygen bubbles on the anode side by the magnetic gradient force.

3.4   Electrodeposition of Metal

3.4.1   Magnetic Stirring by the Lorentz Force

Applying magnetic fields to electrochemical reactions has been intensively studied in the past [24, 25], largely encouraged by enhanced mass transfer found for electrode-parallel magnetic fields. According to Eq. (3.1), electrode-normal oriented current then causes a Lorentz force that drives an electrode-parallel flow, normal to the direction of the magnetic field. The effect on mass transfer can easily be understood from the corresponding enrichment of the boundary layer with bulk electrolyte, thus increasing the diffusional mass flux towards the electrode. Recent progress is based on an improved understanding of the magnetic forcing. Applying homogeneous magnetic fields to cuboid cells with vertical wall electrodes is typically not very successful, as only the small rotational part of the Lorentz force, regardless of the field orientation, creates weak horizontal stirring of the electrolyte [26–28]. Instead, tailored *inhomogeneous* magnetic fields can be used to create strong vertical stirring in the cells. The forced flow counteracts the stratification of the electrolyte caused by solutal buoyancy and is able to increase the homogeneity of the deposit thickness considerably [29, 30]. However, as small electrodes are frequently submerged in larger cells, the flow structure generated in the cell may be quite complex, and it is difficult to draw conclusions based only on the direction of the magnetic field with respect to the electrode, as will be shown below.

3.4.2   Structuring Deposits with the Magnetic Gradient Force

The magnetic gradient force comes into play from the magnetic property of constituents of the electrolyte and the exposure to an inhomogeneous magnetic field. Experiments of Tschulik et al. [31–33] and Dunne et al. [34–36] on the deposition of different metals at electrodes that are magnetically structured on a milli- and micrometer scale delivered results that were surprising at first sight. The deposition rate of paramagnetic ions (e.g. Cu^{2+}) in electrolytes of moderate concentration was found to increase in the vicinity of small permanent magnets or magnetized ferromagnetic elements arranged underneath the cathode. This was surprising, as the magnetic susceptibility of the electrolyte was negative because it was dominated by the diamagnetic water molecules (Eq. 3.3). Adding electrochemically inert and strongly paramagnetic ions (e.g. Mn^{2+}) to the electrolyte turned the former elevation of the deposit into a valley (see Fig. 3.1a, b). Depositing instead diamagnetic ions such as Bi^{3+} did not lead to any local alteration of the deposit thickness in the vicinity of the magnetic elements, which could be understood from the very small magnitude of the magnetic susceptibility of diamagnetic ions in general when compared to paramagnetic ions. However, adding electrochemically inert Mn^{2+} ions locally

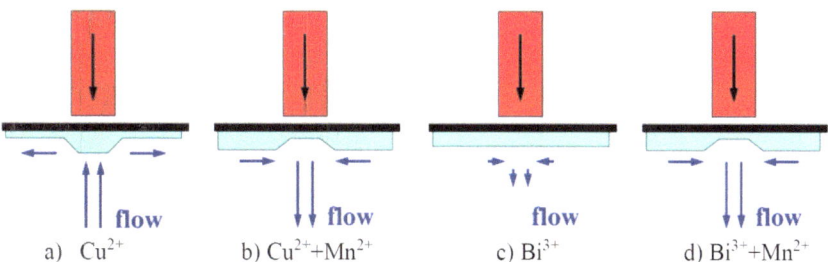

Fig. 3.1 Sketch of the local flow forced near the cathode (black, top position) where a metal (**a, b**: copper, **c, d**: bismuth; all in light blue) is deposited. The electrolyte contains copper or bismuth ions with (**b, d**) or without (**a, c**) manganese ions in excess. The magnetic element above the cathode is drawn in red

reduced the deposition rate, similar to the combined copper case already mentioned (see Fig. 3.1c, d).

Combined fluid-mechanical and electrochemical reasoning is the key to understand these results. As only the rotational part of the magnetic gradient force is able to drive a flow, only the gradient of the magnetic susceptibility of the solution determines the direction of the flow forced [37]. When considering the case of plain copper ions, despite the electrolyte has a negative magnetic susceptibility, its gradient is determined by the concentration gradient of the copper ions only. Thus, a flow towards the electrode is driven by the magnetic gradient force that enriches the boundary layer and therefore enhances the deposition at the magnetic element. Adding electrochemical reasoning, the inverse copper patterning in case of an electrolyte with Mn^{2+} ions can also easily be understood. The key point is that even though the Mn^{2+} ions do not take part in the electrochemical reaction, their concentration near the cathode increases, as electrical neutrality must hold true outside the double layer [38]. As the gradient of the Mn^{2+} concentration dominates the gradient of the solution susceptibility, its sign is opposite compared to the pure copper case, and the direction of flow driven by the magnetic gradient force is reversed. Thus, a wall-parallel flow of depleted electrolyte is approaching the magnetic element where it leaves the electrode, thereby yielding locally a lower deposition rate. The same argument seamlessly explains the case of Bi^{3+} deposition with inert Mn^{2+} ions, as again the inert but strongly paramagnetic ions dominate the susceptibility gradient [39–41]. A broader summary of the reasoning can be found in [42, 43].

The question how tall the structures may become during deposition was addressed only recently. During growth, the surface departs from the region of largest magnetic gradients underneath the cathode, and the structuring effect is reduced. Furthermore, at larger times, solutal buoyancy starts to disturb the structuring effect. It is therefore advantageous to perform the deposition in a pulse-reversed mode to frequently rebuild the concentration boundary layer and retard the action of buoyancy at horizontal electrodes [44].

3.4.3 Nano-Structuring of Metal Layers

The last question addressed here is whether magnetic fields can be beneficial for the manufacturing of nanostructured layers of metals. In electrodeposition, nano-structuring is conventionally achieved by adding capping agents (e.g. Cl^- ions) to the electrolyte to damp or enhance growth in different crystallographic directions [45]. Recently it was shown that electrode-normal magnetic fields force a circumferential flow near conical elevations of metal deposits by action of the Lorentz force. The resulting centrifugal acceleration then gives rise to a secondary downward flow that enhances conical growth [46]. The magnetic gradient force at ferromagnetic cones was found to drive a downward flow as well, thus supporting the conical growth [47]. However, electrodeposition of nano-structured nickel layers in a magnetic field has not yet shown that capping agents can be completely omitted [48]. A scaling analysis performed recently found that the supporting flow driven by the magnetic forces gets weaker as the size of the cone reduces. At the nanoscale, the support of the magnetic gradient force remains substantial, whereas the support of the Lorentz force has decayed [49]. However, experiments to deposit nickel nanostructures did not yet give strong support, as global cell flow driven by the Lorentz force supersedes the beneficial local flow near the cones. Therefore, improved cell and electrode designs are currently under discussion to reduce the global cell flow and thus enable magnetic support for depositing nanostructures [50].

Acknowledgements I am very grateful to the organizers of the MaMI Workshop who gave me the opportunity to present this review in Mittelwhir on June 14, 2022. Only the fruitful cooperation with all the colleagues listed as coauthors in our papers referenced here made this work possible. I am especially grateful to Tom Weier, Gunter Gerbeth, Andreas Bund, Margitta Uhlemann, Kerstin Eckert, Christian Cierpka, Jochen Fröhlich, Kristina Tschulik, Xuegeng Yang, Mengyuan Huang and Piotr Zabinski.

References

1. P.A. Davidson, An Introduction to Magnetohydrodynamics (Cambridge University Press, 2001)
2. R.E. Rosensweig, Ferrohydrodynamics (Cambridge University Press, 1985)
3. J.M.D. Coey, F.M.F. Rhen, P. Dunne, S. McMurry, The concentration gradient force—is it real? J. Solid State Electrochem. 11, 711–717 (2007)
4. M. Suwa, S. Tsukahara, H. Watarai, Application of magnetic force and electromagnetic force in micro-analytical systems, Lab Chip 23, 1097–11,127 (2023)
5. J.M.D. Coey, R. Aogaki, F. Byrne, P. Stamenov, Magnetic stabilization and vorticity in submillimeter paramagnetic liquid tubes. PNAS 106, 8811–8817 (2009)
6. B. Fritzsche, G. Mutschke, T. Meinel, X. Yang, Z. Lei, K. Eckert, Oscillatory surface deformation of paramagnetic rare earth solutions driven by an inhomogeneous magnetic field. Phys. Rev. E 101, 062601 (2020)
7. Z. Lei, B. Fritzsche, R. Salikhov, K. Schwarzenberger, O. Hellwig, K. Eckert, Magnetic separation of rare-earth ions: Property database and Kelvin force distribution. J. Phys. Chem. C 126, 2226–2233 (2022)

8. K. Kolczyk-Siedlecka, M. Wojnicki, X. Yang, G. Mutschke, P. Zabinski, Experiments on the magnetic enrichment or rare-earth metal ions in aqueous solutions in a microflow device. J. Flow Chem. **9**, 175–185 (2019)
9. V. Haehnel, F.Z. Khan, G. Mutschke, C. Cierpka, M. Uhlemann, I. Fritsch, Combining magnetic forces for contactless manipulation of fluids in microelectrode-microfluidic systems. Sci. Rep. **9**, 5103 (2019)
10. R. Aogaki, K. Fueki, T. Mukaibo, T., Application of magnetohydrodynamic effect to the analysis of electrochemical reactions. Denki Kagaku 43, 504–508 and 509–514 (1975)
11. Y. Sasakawa, S. Takezawa, Y. Sugawara, Y. Kyotani, The superconducting MHD-propelled ship YAMATO-1. In: Proceeding of the 4th international conference and exhibition: world congress on superconductivity, vol. 1 (1995), pp. 167–176
12. C.-S. Chan, J.-H. Cheng, C.-H. Zeng, J.-R. Huang, Y.-H. Chen, Y.-J. Chen, T.T. Pham, W.-H. Chao, J.-T. Jeng, Y.-L. Liu, K.-C. Pan, Y.-H. Li, C.-Y. Chen, Design of marine vehicle powered by magnetohydrodynamic thruster. Magnetohydrodynamics **56**, 51–65 (2020)
13. A. Gailitis, O. Lielausis, On the possibility to reduce the hydrodynamical resistance of a plate in an electrolyte. Appl. Magnetohydrodyn. Rep. Riga Inst. Phys. **12**, 143–146 (1961). ((in Russian))
14. T. Albrecht, R. Grundmann, G. Mutschke, G. Gerbeth, On the stability of the boundary layer subject to a wall-parallel Lorentz force. Phys. Fluids **18**, 098103 (2006)
15. T. Albrecht, H. Metzkes, R. Grundmann, G. Mutschke, G. Gerbeth, Tollmien-Schlichting wave damping by a streamwise oscillating Lorentz force. Magnetohydrodynamics **44**(3), 205–222 (2008)
16. T. Weier, G. Gerbeth, G. Mutschke, O. Lielausis, G. Lammers, Control of flow separation using electromagnetic forces. Flow Turbul. Combust. **71**, 5–17 (2003)
17. G. Mutschke, G. Gerbeth, T. Albrecht, R. Grundmann, Separation control at hydrofoils using Lorentz forces. Eur. J. Mech. B/Fluids **25**, 137–152 (2006)
18. T. Albrecht, J. Stiller, H. Metzkes, T. Weier, G. Gerbeth, Electromagnetic flow control in poor conductors. Eur. Phys. J. Special Topics **220**, 275–285 (2013)
19. D. Pletcher, X. Li, Prospects for alkaline zero gap water electrolysers for hydrogen production. Int. J. Hydrgen En. **36**, 15089–15104 (2011)
20. J.A. Koza, S. Mühlenhoff, P. Zabinski, P.A. Nikrityuk, K. Eckert, M. Uhlemann, A. Gebert, T. Weier, L. Schultz, S. Odenbach, Hydrogen evolution under the influence of a magnetic field, Electrochimica Acta 56:2665–2675 (2011)
21. D. Baczyzmalski, F. Karnbach, G. Mutschke, X. Yang, K. Eckert, M. Uhlemann, C. Cierpka, Growth and detachment of single hydrogen bubbles in a magnetohydrodynamic shear flow. Phys. Rev. Fluids **2**, 093701 (2017)
22. D. Baczyzmalski, F. Karnbach, X. Yang, G. Mutschke, M. Uhlemann, K. Eckert, C. Cierpka, On the electrolyte convection around a hydrogen bubble evolving at a microelectrode under the influence of a magnetic field. J. Electrochem. Soc. **163**, E248–E257 (2016)
23. G. Mutschke, J. Fröhlich, X. Yang, K. Eckert, F. Karnbach, M. Uhlemann, D. Baczyzmalski, C. Cierpka, Numerical simulation of mass transfer and convection near a hydrogen bubble during water electrolysis in a magnetic field. Magnetohydrodynamics **53**, 193–199 (2017)
24. T.Z. Fahidy, The effect of magnetic fields on electrochemical processes, in *Modern Aspects of Electrochemistry*, ed. by B.E. Conway. et al. no, vol. 32 (Kluver/Plenum press, New York, 1999), pp.333–353
25. A. Bund, A. Ispas, G. Mutschke, Magnetic field effects on electrochemical metal depositions. Sci. Technol. Adv. Mater. 9 (2008) 024208
26. G. Mutschke, A. Bund, On the 3D character of the magnetohydrodynamic effect during metal electrodeposition in cuboid cells. Electrochem. Comm. **10**, 597–601 (2008)
27. G. Mutschke, A. Hess, A. Bund, J. Fröhlich, On the origin of horizontal counter-rotating electrolyte flow during copper magnetoelectrolysis. Electrochim. Acta **55**, 1543–1547 (2010)
28. D. Koschichow, G. Mutschke, X. Yang, A. Bund, J. Fröhlich, Numerical simulation of the onset of mass transfer and convection in copper electrolysis subjected to a magnetic field. Russ. J. Electrochem. **48**, 756–765 (2012)

29. S. Mühlenhoff, G. Mutschke, D. Koschichow, X. Yang, A. Bund, S. Odenbach, K. Eckert, Lorentz-force-driven convection during copper magnetoelectrolysis in the presence of a supporting buoyancy force. Electrochim. Acta **69**, 209–219 (2012)
30. S. Mühlenhoff, G. Mutschke, M. Uhlemann, X. Yang, S. Odenbach, J. Fröhlich, K. Eckert, On the homogenization of the thickness of Cu deposits by means of MHD convection within small dimension cells. Electrochem. Comm. **36**, 80–83 (2013)
31. K. Tschulik, J.A. Koza, M. Uhlemann, A. Gebert, L. Schultz, Effects of well-defined magnetic field gradients on the electrodeposition of copper and bismuth. Electrochem. Comm. **11**, 2244 (2009)
32. K. Tschulik, C. Cierpka, G. Mutschke, A. Gebert, L. Schultz, M. Uhlemann, Clarifying the mechanism of reverse-structuring during electrodeposition in magnetic gradient fields. Anal. Chem. **84**, 2328–2334 (2011)
33. K. Tschulik, X. Yang, G. Mutschke, M. Uhlemann, K. Eckert, R. Süptitz, L. Schultz, A. Gebert, How to obtain structured metal deposits from diamagnetic ions in magnetic gradient fields? Electrochem. Comm. **13**, 946–950 (2011)
34. P. Dunne, L. Massa, J.M.D. Coey, Magnetic structuring of electrodeposits, Phys. Rev. Lett. 107 (2011) 024501
35. P. Dunne, R. Soucaille, K. Ackland, J.M.D. Coey, Magnetic structuring of linear copper electrodeposits. J. Appl. Phys. **111**, 07B915 (2012)
36. P. Dunne, J.M.D. Coey, Patterning metallic electrodeposits with magnet arrays, Phys. Rev. B 85 (2012) 224411
37. G. Mutschke, K. Tschulik, T. Weier, M. Uhlemann, A. Bund, J. Fröhlich, On the action of magnetic gradient forces in micro-structured copper deposition. Electrochim. Acta **55**, 9060–9066 (2010)
38. C. Wagner, The role of natural convection in electrolytic processes. J. Electrochem. Soc. **95**, 161–173 (1949)
39. G. Mutschke, K. Tschulik, M. Uhlemann, A. Bund, J. Fröhlich, Comment on "Magnetic structuring of electrodeposits." Phys. Rev. Lett. **109**, 229401 (2012)
40. G. Mutschke, K. Tschulik, T. Weier, M. Uhlemann, A. Bund, A. Alemany, J. Fröhlich, On the action of magnetic gradient forces in micro-structured copper deposition. Magnetohydrodynamics **48**, 299–304 (2012)
41. G. Mutschke, K. Tschulik, M. Uhlemann, J. Fröhlich, Numerical simulation of the mass transfer of magnetic species at electrodes exposed to small-scale gradients of the magnetic field. Magneto-hydrodynamics **51**, 369–374 (2015)
42. G. Mutschke, Über den Einfluss von Magnetfeldern auf Strömung und Stoffübergang bei der elektrochemischen Metallabscheidung. TUDpress, Dresden (2013). ISBN 978-3-944331-40-9
43. M. Uhlemann, K. Tschulik, A. Gebert, G. Mutschke, J. Fröhlich, A. Bund, X. Yang, K. Eckert, Structured electrodeposition in magnetic gradient fields. Eur. Phys. J. Spec. Top. **220**, 287–302 (2013)
44. M. Huang, M. Uhlemann, K. Eckert, G. Mutschke, Pulse reverse plating of copper microstructures in magnetic gradient fields. Magnetoelectrochemistry **8**, 66 (2022)
45. T. Hang, M. Li, Q. Fei, D. Mao, Characterization of nickel nanocones routed by electrodeposition without any template. Nanotechnology **19**, 035201 (2008)
46. M. Huang, G. Marinaro, X. Yang, B. Fritzsche, Z. Lei, M. Uhlemann, K. Eckert, G. Mutschke, Mass transfer and electrolyte flow during electrodeposition on a conically shaped electrode under the influence of a magnetic field. J. Electroanalyt. Chem. 842, 203–213 (2019)
47. M. Huang, K. Eckert, G. Mutschke, Magnetic-field-assisted electrodeposition of metal to obtain conically structured ferromagnetic layers. Electrochim. Acta **365**, 137374 (2021)
48. K. Skibinska, M. Huang, G. Mutschke, K. Eckert, G. Wloch, M. Wojnicki, P. Zabinski, On the electrodeposition of conically nano-structured nickel layers assisted by a capping agent. J. Electroanalyt. Chem. **904**, 115935 (2022)
49. M. Huang, K. Skibinska, P. Zabinski, M. Wojnicki, G. Wloch, K. Eckert, G. Mutschke, On the prospects of magnetic-field-assisted electrodeposition of nano-structured ferromagnetic layers. Electrochimica Acta 420, 140422 (2022)

50. M. Huang, K. Skibinska, P. Zabinski, K. Eckert, G. Mutschke, Minimizing global cell flow to support the electrodeposition of nano-structured metal layers in a magnetic field. Magnetohydrodynamics **58**, 339–348 (2022)

Chapter 4
Phenomenological Models of Magnetizable Fluids

Andrejs Cēbers

In the description of the motion of magnetic liquids (ferrofluids) a principal issue is the acting stress. It is derived in different textbooks, including the well known book by Rosensweig [1], where the ponderomotive forces are derived by the energetical approach, and many others [2–4]. In this article a general approach that considers the conservation laws and is also valid in non-equilibrium situations is reviewed. It is based on an extension of an overview of the description of magnetizable media in [5].

One difficulty in describing magnetizable media is deciding what should be taken as the energy of the electromagnetic field and what as the energy of the medium. The problem is that in an applied field the energy of the medium depends on the field and a simple distinction between the two energies is not evident. Here we make our choice based on the Maxwell equations in continous media, which in a quasistationary approximation read

$$\nabla \times \vec{E} = -\frac{1}{c}\frac{\partial \vec{B}}{\partial t}; \ \nabla \times \vec{H} = \frac{4\pi}{c}\vec{j} \tag{4.1}$$

as usually

$$\vec{B} = \vec{H} + 4\pi \vec{M}; \ \nabla \cdot \vec{B} = 0. \tag{4.2}$$

Equation (4.1) give

$$\frac{1}{4\pi}\vec{H}\frac{\partial \vec{B}}{\partial t} + \vec{E} \cdot \vec{j} + \frac{c}{4\pi}\nabla \cdot [\vec{E} \times \vec{H}] = 0, \tag{4.3}$$

which using the identity $\varrho\frac{d\tilde{a}}{dt} = \frac{\partial a}{\partial t} + \nabla \cdot (\varrho\tilde{a}\vec{v})$ ($\tilde{a} = a/\varrho$ is the quantity per unit mass) may be rewritten as follows

A. Cēbers (✉)
Department of Physics, MMML Lab, University of Latvia, Jelgavas-3, Riga LV-1004, Latvia
e-mail: andrejs.cebers@lu.lv

© The Author(s) 2024
B. Doudin et al. (eds.), *Magnetic Microhydrodynamics*, Topics in Applied Physics 120,
https://doi.org/10.1007/978-3-031-58376-6_4

$$\varrho\frac{d}{dt}\frac{\vec{H}^2}{8\pi\varrho} - \nabla\cdot(\frac{\vec{H}^2}{8\pi}\vec{v}) + \vec{H}\varrho\frac{d\tilde{M}}{dt} - \vec{H}\partial_i(v_i\vec{M}) + \vec{E}\cdot\vec{j} + \frac{c}{4\pi}\nabla\cdot[\vec{E}\times\vec{H}] = 0.$$

(4.4)

Equation (4.4) shows that the volume density of the electromagnetic field energy may be identified as $e_f = \vec{H}^2/(8\pi)$. Other terms in Eq.(4.4) may be identified as source terms for the internal and mechanical energy of the medium and as a flux of the electromagnetic energy.

For the total energy of the system the following local conservation law is valid (for an overview of energy conservation in systems interacting with electromagnetic field see [6])

$$\varrho\frac{d}{dt}\left(\frac{\vec{v}^2}{2} + \tilde{e}_f + \tilde{e}\right) = \partial_k(v_i(\sigma_{ik} + T_{ik})) - \nabla\cdot\vec{j}_f - \nabla\cdot\vec{j}_q,$$

(4.5)

where the first term on the right side in Eq.(4.5) corresponds to the work done by the total stress $\sigma_{ik} + T_{ik}$ (T_{ik} is the electromagnetic field contribution to the stress identified further), the second term describes the electromagnetic energy flux and the third describes the transfer of heat. Similar conservation laws are valid for momentum and mass

$$\varrho\frac{dv_i}{dt} = \partial_k(\sigma_{ik} + T_{ik});$$

(4.6)

$$\varrho\frac{d}{dt}\frac{1}{\varrho} = \nabla\cdot\vec{v}.$$

(4.7)

It may be noted that (4.6) in the case of equilibrium magnetization and incompressible liquid reads

$$\varrho\frac{d\vec{v}}{dt} = -\nabla p + \eta\Delta\vec{v} + (\vec{M}\cdot\nabla)\vec{H} + \frac{1}{c}[\vec{j}\times\vec{H}].$$

(4.8)

We see that in term for the Lorenz force on current we have $\vec{H}$ and not $\vec{B}$ as was pointed out by A.Einstein et al. in [7]. In [8] it is shown that electromagnetic force in Eq.(4.8) is reduced to $\frac{1}{c}[\vec{j}\times\vec{B}]$ if the condition $\frac{\partial H_i}{\partial x_j} = 0$ is valid. Taking $\vec{j}_f = \frac{c}{4\pi}[\vec{E}'\times\vec{H}]$, where $\vec{E}' = \vec{E} + \frac{1}{c}[\vec{v}\times\vec{B}]$ is the electric field strength in the reference frame of the material element, relation (4.4) may be put in the following form

$$\varrho\frac{d}{dt}\frac{\vec{H}^2}{8\pi\varrho} + \vec{H}\varrho\frac{d\tilde{M}}{dt} - \partial_k\left(v_i\left(\frac{H_iB_k}{4\pi} - \frac{\vec{H}^2}{8\pi}\right)\right) + v_i\vec{M}\partial_i(\vec{H})$$
$$+ \vec{E}\cdot\vec{j} + \frac{c}{4\pi}\nabla\cdot[\vec{E}'\times\vec{H}] = 0.$$

(4.9)

The equation for the kinetic energy reads

$$\varrho\frac{d}{dt}\frac{\vec{v}^2}{2} = \partial_k(v_i\sigma_{ik}) - \sigma_{ik}\frac{\partial v_i}{\partial x_k} + v_i\partial_k T_{ik}.$$

(4.10)

Subtraction of $\varrho \frac{d}{dt} \frac{\vec{H}^2}{8\pi\varrho} + \varrho \frac{d}{dt} \frac{\vec{v}^2}{2}$ given by Eqs. (4.9, 4.10) from Eq. (4.5) and identifying $T_{ik} = \frac{H_i B_k}{4\pi} - \frac{H^2}{8\pi} \delta_{ik}$ allows us to obtain the source term for the internal energy of media $\tilde{e}$ which reads

$$\varrho \frac{d\tilde{e}}{dt} = \varrho \vec{H} \cdot \frac{d\tilde{M}}{dt} + v_i \vec{M} \partial_i \vec{H} + \sigma_{ik} \frac{\partial v_i}{\partial x_k} + \vec{E} \cdot \vec{j} - v_i \partial_k T_{ik} - \nabla \cdot \vec{j}_q. \qquad (4.11)$$

Using $\partial_i H_k - \partial_k H_i = e_{ikj} \frac{4\pi}{c} j_j$ and

$$- v_i \partial_k T_{ik} + v_i M_k \partial_i H_k = \frac{1}{c} [\vec{v} \times \vec{B}] \cdot \vec{j}$$

the equation for the internal energy reads

$$\varrho \frac{d\tilde{e}}{dt} = \varrho \vec{H} \cdot \frac{d\tilde{M}}{dt} + \sigma_{ik} \frac{\partial v_i}{\partial x_k} + \vec{E}' \cdot \vec{j} - \nabla \cdot \vec{j}_q. \qquad (4.12)$$

Taking the stress tensor of the medium as $\sigma_{ik} = -p\delta_{ik} + \tau_{ik}$, where τ_{ik} is the viscous stress tensor, we have for the case when the magnetization is in thermal equilibrium the following expression

$$\varrho \frac{d\tilde{e}}{dt} = \varrho T \frac{d\tilde{S}}{dt} - p\varrho \frac{d}{dt} \frac{1}{\varrho} + \varrho \vec{H} \cdot \frac{d\tilde{M}}{dt}, \qquad (4.13)$$

where for the specific entropy $\tilde{S}$ we have

$$\varrho T \frac{d\tilde{S}}{dt} = \tau_{ik} \frac{\partial v_i}{\partial x_k} + \vec{E}' \cdot \vec{j} - \nabla \cdot \vec{j}_q. \qquad (4.14)$$

According to the relation (4.13) the internal energy is defined by the equation of state $\tilde{e} = \tilde{e}(\tilde{S}, \varrho, \tilde{M})$. In the case when the magnetization is in a non-equilibrium state the relation (4.13) is put in the following form [5]

$$\varrho \frac{d\tilde{e}}{dt} = \varrho T \frac{d\tilde{S}}{dt} - p\varrho \frac{d}{dt} \frac{1}{\varrho} + \varrho \vec{H}_e \cdot \frac{d\tilde{M}}{dt}. \qquad (4.15)$$

In (4.15) instead of $\vec{H}$ the relation for the internal energy contains the effective field $\vec{H}_e$ according to which the magnetization is given by the equilibrium magnetization law $\vec{M} = \vec{M}_{eq}(\vec{H}_e)$. The effective field was introduced in [9] in order to describe a non-equilibrium state of the magnetization for diluted ferrofluids. As a result for the entropy production σ we obtain

$$\varrho \frac{d\tilde{S}}{dt} = -\nabla \cdot \vec{j}_s + \sigma, \qquad (4.16)$$

where

$$T\sigma = \vec{j}_q \cdot \nabla \frac{1}{T} + \tau_{ik}\frac{\partial v_i}{\partial x_k} + \vec{E}' \cdot \vec{j} - (\vec{H}_e - \vec{H})\rho\frac{d\tilde{M}}{dt}. \tag{4.17}$$

On the basis of the relation (4.17) for the entropy production the linear phenomenological laws for heat transfer, viscous stress, the charge transfer and magnetic relaxation may be formulated.

Let us consider in detail the phenomenology of the magnetic relaxation. The total stress tensor as it follows from the requirement of the angular momentum conservation is symmetric $\sigma_{ik} + T_{ik} = \sigma_{ki} + T_{ki}$ (we are neglecting the internal angular momentum due to the spinning of particles). As a result the term $\tau_{ik}\frac{\partial v_i}{\partial x_k}$ may be transformed as follows

$$\tau_{ik}\frac{\partial v_i}{\partial x_k} = \tau_{ik}^s \frac{1}{2}\left(\frac{\partial v_i}{\partial x_k} + \frac{\partial v_k}{\partial x_i}\right) + \tau_{ik}^a \frac{1}{2}\left(\frac{\partial v_i}{\partial x_k} - \frac{\partial v_k}{\partial x_i}\right).$$

Since $\tau_{ik}^a = -T_{ik}^a$ and $T_{ik}^a = -\frac{1}{2}e_{ikl}[\vec{M} \times \vec{H}]_l$ we have

$$\tau_{ik}^a \frac{1}{2}\left(\frac{\partial v_i}{\partial x_k} - \frac{\partial v_k}{\partial x_i}\right) = [\vec{M} \times \vec{H}] \cdot \vec{\Omega}_0,$$

where $\vec{\Omega}_0 = \frac{1}{2}\nabla \times \vec{v}$ is the angular velocity of the local rotation of the fluid. According to this the entropy production may be put in the following form ($\vec{M} \parallel \vec{H}_e$)

$$T\sigma = \vec{j}_q \cdot \nabla \frac{1}{T} + \tau_{ik}\frac{\partial v_i}{\partial x_k} + \vec{E}' \cdot \vec{j} - (\vec{H}_e - \vec{H})\left(\rho\frac{d\tilde{M}}{dt} - [\vec{\Omega}_0 \times \vec{M}]\right). \tag{4.18}$$

We see that as a thermodynamic flux for the magnetization relaxation there appears $\rho\frac{d\tilde{M}}{dt} - [\vec{\Omega}_0 \times \vec{M}]$, which describes the magnetization relaxation in the reference frame of a rotating material element. Accounting for this term allows one to describe such effects as the increase of the effective viscosity of the ferrofluid in an applied field [10].

The general approach according to which the magnetic relaxation equation is formulated, in our opinion, resolves the issue of the correct form of this equation which was under dispute in the literature [11–14]. The kinetic coefficients for the magnetic relaxation are obtained considering the Brownian motion of magnetic dipoles in the effective field approximation [9] and read as follows

$$\left(\rho\frac{d\tilde{M}}{dt} - [\vec{\Omega}_0 \times \vec{M}]\right)_{\parallel,\perp} = -\frac{1}{\gamma_{\parallel,\perp}}\left(\vec{H}_e - \vec{H}\right)_{\parallel,\perp}, \tag{4.19}$$

where

$$\gamma_{\parallel}^{-1} = \frac{nm^2}{\alpha} \frac{2L(\xi_e)}{\xi_e}; \ \gamma_{\perp}^{-1} = \frac{nm^2}{\alpha} \left(1 - \frac{L(\xi_e)}{\xi_e}\right). \tag{4.20}$$

Here m is the magnetic moment of a colloidal particle, n is their concentration, α is the rotational drag coefficient per unit volume, ξ_e is the Langevin parameter of the effective field $\xi_e = mH_e/k_BT$ and $L(\xi_e)$ is the Langevin function.

The magnetic relaxation Eq. (4.19) and the equation of motion of an incompressible ferrofluid (4.6) (the Lorentz force on the electric current is neglected) give the closed set of equations describing its motion in the case of the non-equilibrium magnetization

$$\varrho \frac{d\vec{v}}{dt} = -\nabla p + \eta \Delta \vec{v} + (\vec{M} \cdot \nabla)\vec{H} + \frac{1}{2}\nabla \times [\vec{M} \times \vec{H}]. \tag{4.21}$$

We draw attention to the appearance of the volume force $\frac{1}{2}\nabla \times [\vec{M} \times \vec{H}]$ in (4.21) due to the inhomogeneity of the volume torque $[\vec{M} \times \vec{H}]$. The set of equations (4.19,4.21) describes a broad variety of phenomena connected with the spinning of magnetic particles as edge flows of a ferrofluid with free boundaries under the action of a rotating field [15], the increase of its effective viscosity in an applied field [10] and others. The expression for the dissipative function allows one to estimate the effect of magnetic hyperthermia which is of great interest due to its applications in biomedicine [16].

In the case when the magnetization of the ferrofluid is in thermodynamic equilibrium $\vec{M} = M_{eq}(H)\vec{H}/H$ the equation of motion includes the Kelvin force $M\nabla H$ and describes a variety of phenomena, for example, the deformation of droplets in an applied field [17], labyrinthine instabilities [18, 19] and magnetic microconvection [20] among others.

Acknowledgements Author acknowledges support by the grant of the Latvia Council of Science lzp-2020/1-0149 and the European Union's Horizont 2020 project MAMI No.766007.

References

1. R.E. Rosensweig, *Ferrohydrodynamics* (Cambridge University Press, 1985)
2. J.A. Stratton, *Electromagnetic Theory* (MbGraw-Hill Book Company, New York, London, 1941)
3. J.D. Jackson, *Classical Electrodynamics* (John Wiley@Sons, USA, 1999)
4. L.D. Landau, E.M. Lifshitz, *Electrodynamics of Continous Media* (Pergamon Press, Oxford, 1984)
5. E. Blums, A. Cebers, M.M. Maiorov, *Magnetic Liquids* (W de G. Gruyter, Berlin, New York, 1997)
6. P. Penfield, H.A. Haus, *Electrodynamics of Moving Media* (Press, M.I.T, 1967)
7. A. Einstein, J. Laub. Ann. Phys. **26**, 541–550 (1908)
8. K.J. Webb, Phys. Rev. B. 94m064203 (2016)

9. M.A. Martsenyuk, Y.L. Raikher, M.I. Shliomis, On the kinetics of magnetization of suspensions of ferromagnetic particles. Sov. Phys. JETP. **38**, 413–416 (1974)
10. J.P. McTague, Magnetoviscosity of magneticcolloids. J. Chem. Phys. **51**, 133–136 (1969)
11. H.W. Muller, M. Liu, Structure of ferrofluid dynamics. Phys. Rev. E **64**, 061405 (2001)
12. M.I. Shliomis, Comment on "Magnetoviscosity and relaxation in ferrofluids". Phys. Rev. E **64**, 063501 (2001)
13. M.I. Shliomis, Comment on "Structure of ferrofluid dynamics". Phys. Rev. E **67**, 043201 (2003)
14. A. Fang, Consistent hydrodynamics of ferrofluids. Phys. Fluids **34**, 013319 (2022)
15. A. Tsebers, Interfacial stresses in the hydrodynamics of liquids with internal rotations. Magnetohydrodynamics **11**, 63–65 (1975)
16. H. Fatima, T. Charinpaitkul, K.S. Kim, Fundamentals to apply magnetic nanoparticles for hyperthermia therapy. Nanomaterials **11**, 1203 (2021)
17. J. Bacri, D. Salin, Instability of ferrofluid magnetic drops under magnetic field. J. Phys. Lett. **43**, 649–654 (1982)
18. A. Tsebers, M.M. Mayorov, Magnetostatic instabilities in plane layers of magnetizable fluids. Magnetohydrodynamics **16**, 21–27 (1980)
19. A. Tsebers, M.M. Mayorov, Structures of interface a bubble and magnetic fluid in a field. Magnetohydrodynamics **16**, 231–235 (1980)
20. G. Kitenbergs, A. Tatulcenkovs, K. Erglis, O. Petrichenko, R. Perzynski, A. Cebers, Magnetic field driven micro-convection in the Hele-Shaw cell: the Brinkman model and its comparison with experiment. J. Fluid Mech. **774**, 170–191 (2015)

Part II
Movers and Shakers

Chapter 5
Scaling and Flow Profiles in Magnetically Confined Liquid-In-Liquid Channels

Arvind Arun Dev, Florencia Sacarelli, G. Bagheri, Aleena Joseph, Anna Oleshkevych, E. Bodenschatz, Peter Dunne, Thomas Hermans, and Bernard Doudin

5.1 Introduction

In the realm of fluid mechanics, there is a need for new approaches to diminish friction in fluid flow [1–5] and therefore limit the transport energy loss [6, 7]. A low-friction environment can exist under exotic conditions like superfluidity at low temperature [8], nanofluidic channels (nanopores) made up of atomically flat crystals [9], or superfluid-like behaviour of bacterial suspensions [10]. These specific physical conditions allow a velocity flow profile that approaches plug flow with a

A. A. Dev · A. Joseph · A. Oleshkevych · P. Dunne · B. Doudin (✉)
Université de Strasbourg, CNRS, IPCMS UMR 7504, 23 Rue du Loess, 67034 Strasbourg, France
e-mail: bernard.doudin@ipcms.unistra.fr

A. Joseph
e-mail: aleena.joseph@ipcms.unistra.fr

P. Dunne
e-mail: peter.dunne@ipcms.unistra.fr

A. A. Dev (✉)
Laboratoire Colloïdes et Materiaux Divises, Chemistry Biology & Innovation, ESPCI, PSL Research University, Paris 75005, France
e-mail: arvind.dev@espci.psl.eu; dev.arvinda@gmail.com

School of Applied and Engineering Physics, Cornell University, Ithaca, NY 14853, USA

F. Sacarelli · G. Bagheri · T. Hermans (✉)
Université de Strasbourg, CNRS UMR 7140, 4 Rue Blaise Pascal, 67081 Strasbourg, France
e-mail: thomas.hermans@imdea.org

E. Bodenschatz
Laboratory for Fluid Physics, Pattern Formation and Biocomplexity, Max Planck Institute for Dynamics and Self-Organization, 37077 Göttingen, Germany
e-mail: eberhard.bodenschatz@ds.mpg.de

© The Author(s) 2024
B. Doudin et al. (eds.), *Magnetic Microhydrodynamics*, Topics in Applied Physics 120,
https://doi.org/10.1007/978-3-031-58376-6_5

constant velocity, in stark contrast to the expected parabolic behaviour predicted by Poiseuille's flow profile constrained by no-slip boundary conditions at solid walls. However, a room-temperature system with low wall friction and thus low shear to transport delicate particles or cells is rare and feasible only with the use of lubricants in coaxial flows [11], or infused in channel surfaces [11–13]. The use of such lubricating layers is restricted by their limited robustness, arising from drainage of the lubricant or the occurrence of hydrodynamic instabilities [14–16]. Pioneering work in the 80s demonstrated friction reduction by trapping ferrofluid lubricants inside large pipes using magnetic forces [17, 18]. In simple terms, the magnetic force holds a layer of ferrofluid in place, over which the transported liquid flows. Thus, the absence of the solid wall of the pipe leads to reduction of friction. Here we review how optimization of magnetic force fields can stabilise small fluidic tubes, aiming to scale fluidic circuitry down to micro- and nanofluidic sizes. Shear forces, which are a key bottleneck for handling delicate biological objects [19], are minimised by this approach. It therefore opens the possibility to realize robust pressure-controlled microfluidic devices down to the smallest sizes.

This work is based on our initial findings of the use of a quadrupolar field to stabilize microfluidic channels and the observed large reduction of friction [20, 21]. Our aim is two-fold: first specify the physical properties that govern size and friction of a transported fluid and then provide direct experimental insight by imaging a fluidic circuit and its velocity profile characteristics. We show that channels of size below 10 μm can be obtained and a nearly constant (plug-)flow velocity profile can be reached in this liquid-in-liquid design. In the last section, we propose insights into a novel fluidic behaviour that can be observed when the cylindrical symmetry of the system is perturbed.

5.2 Magnetic Force Field Design

In the presence of a non-uniform magnetic field H, a material with uniform magnetization M, small enough to neglect its generated demagnetizing field, experiences a Kelvin force:

Institute for Dynamics of Complex Systems, University of Göttingen, 37077 Göttingen, Germany

Laboratory of Atomic and Solid State Physics, Cornell University, Ithaca, NY 14853, USA

Sibley School of Mechanical and Aerospace Engineering, Cornell University, Ithaca, NY 14853, USA

F. Sacarelli
e-mail: sacarelli@unistra.fr

G. Bagheri
e-mail: gholamhossein.bagheri@ds.mpg.de

T. Hermans
IMDEA Nanociencia, C/ Faraday 9, 28049 Madrid, Spain

$$F = \mu_0(\boldsymbol{M} \cdot \boldsymbol{\Delta})\boldsymbol{H} \tag{5.1}$$

where μ_0 is the permeability of free space. The direction of this force depends on the magnetic susceptibility of the material, negative for diamagnetic materials and positive for paramagnetic or ferromagnetic ones, i.e. diamagnets are repulsed from regions of highest magnetic field, while the reverse is true for paramagnets and ferromagnets.

All the fluidic circuits presented here are based on the use of a pseudo-quadrupolar magnetic field source generated by commercially available N42-grade NdFeB permanent magnets with remanent magnetization $\mu_0 M_r = 1.2$ T. These magnets are arranged to generate a negligible field at the geometric centre of the assembly while radially increasing outward. Figure 5.1 shows cross sections of the two magnetic designs used for experiments, with the white arrows indicating the direction of magnetization of the magnets. Figure 5.1a illustrates the generated nearly axisymmetric magnetic field [20, 21] which in cylindrical coordinates can be approximated by a magnetic field H with a constant gradient:

$$\frac{\partial H(r)}{\partial r} = \frac{4M_r}{\pi w} \tag{5.2}$$

where the magnetic stray field H is related to B via $B = \mu_0 H$, r is the radial distance from the centre, and W the gap between magnets with the same magnetisation (Fig. 5.1). When a paramagnetic or superparamagnetic liquid (i.e. a ferrofluid) is inserted in the region between magnets (Fig. 5.1a), it is attracted to the high field regions, displacing and encapsulating any diamagnetic or weaker paramagnetic liquids which are then confined to the low-field regions. The resulting liquid-in-liquid tube, or '*antitube*' [21], adopts a nearly circular magnetic cross-section.

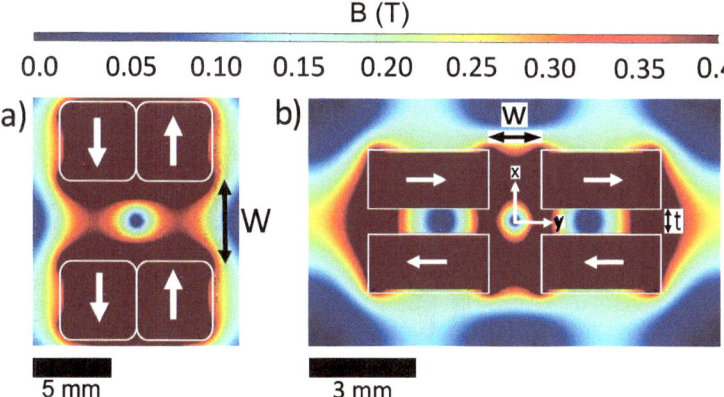

Fig. 5.1 Magnetic design cross section: **a** Magnetic field generated by a four-magnet arrangement with white arrows showing their magnetization direction, $W = 6$ mm. The flow is in z direction (normal to the plots) **b** Magnetic arrangement for the velocimetry experiments, with $W = 1.5$ mm, $t = 0.7$ mm

5.3 The Static Case (No Flow)

At first let us consider a system without flow to test the hypothesis of a cylindrical enclosure for the transport of the liquid of interest and to determine the smallest possible diameter that can be reached. The strong optical absorption of ferrofluids complicates imaging, making X-ray imaging the most appropriate tool, with synchrotron beamline facilities being necessary for micron-range resolution. Figure 5.2 shows the experimental X-ray absorption contrast image using 2D radiography slices and 3D reconstruction from synchrotron tomography imaging. Figure 5.2a reveals the antitube as a brighter region in the centre, of average diameter 80 ± 2 μm, surrounded by the darker ferrofluid, with Fig. 5.2b showing the circular cross-section of the antitube in a vertical slice. Figure 5.2c confirms the cylindrical geometry in the reconstructed data, where the antitube and surrounding ferrofluids are coloured in yellow and blue respectively.

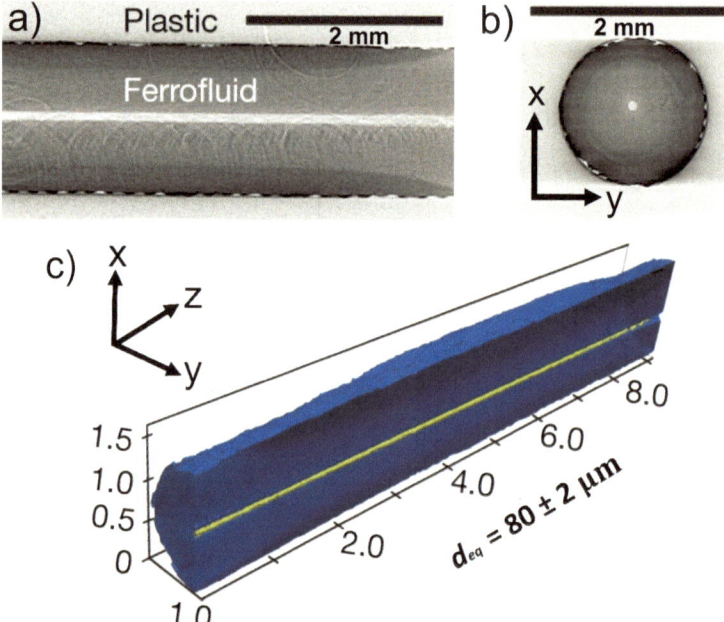

Fig. 5.2 Imaging of the encapsulated antitube. **a** X-ray absorption contrast image, 2D radiography of an antitube [21], the brighter central part is water encapsulated by darker surrounding ferrofluid, **b** cross section of Fig. 5.2a where the brighter central part is the transported liquid surrounded by a darker ferrofluid. **c** Synchrotron X-ray 3D tomographic reconstruction of a water antitube (yellow) with diameter 80 ± 2 μm, surrounded by ferrofluid (blue), the axis units are in mm. Reproduced with permission from [21]

In the static case, the balance of Laplace pressure (i.e. surface tension), magnetic pressure, and magnetic normal traction at the magnetic/non-magnetic interface results in a minimum/equilibrium antitube diameter, d_{eq}. To begin with, the Bernoulli equation with magnetic pressure is given by:

$$P' + \rho g h + \tfrac{1}{2}\rho u^2 - P_m = \text{constant} \tag{5.3}$$

Since the magnetic force is considerably larger than the gravitational force, and we are in static case, we neglect the gravity and velocity terms giving:

$$P' - P_m = \text{constant} \tag{5.4}$$

Eq. 5.4 is the governing equation valid at all locations, where P' is the local pressure and $P_m = \int_0^H \mu_0 M dH = \mu_0 \overline{M} H$ is the fluid magnetic pressure, M is the magnetization of the ferrofluid, $\overline{M}$ is the field averaged magnetisation of the ferrofluid under external magnetic field H. A general boundary condition to solve Eq. 5.4 is:

$$P' + P_n = P_0 + P_c \tag{5.5}$$

where P_0 is atmospheric pressure. $P_n = \tfrac{1}{2}\mu_0 M_I^2$ and $P_c = \tfrac{\sigma}{R}$ are the magnetic normal traction and Laplace pressure respectively, σ is the interfacial tension, and R the radius of the antitube. P_n and P_c act at the magnetic–nonmagnetic interface and at the interface with a nonzero interfacial tension respectively. In our work, both act at the boundary between antitube and ferrofluid. Hence, at the centre of the antitube (location 1), $P_1' = P_0$ and at the magnetic-nonmagnetic interface (location 2), $P_2' = P_0 + P_c - P_n$. Since Eq. 5.4 is valid at all the locations we can write:

$$P_1' - P_{m_1} = P_2' - P_{m_2} \tag{5.6}$$

as at centre of the antitube, $H = 0$, hence $P_{m_1} = 0$ and Eq. 5.6 results in:

$$0 = \tfrac{\sigma}{R} - \mu_0 \overline{M} H_I - \tfrac{1}{2}\mu_0 M_I^2 \tag{5.7}$$

that depicts the balance between surface tension (Laplace pressure), fluid magnetic pressure and the magnetic normal traction, where σ, H_I and M_I are surface tension, magnetic field and ferrofluid magnetisation at the magnetic-nonmagnetic interface respectively. Rearranging Eq. 5.7 gives the equilibrium diameter of the antitube ($d_{eq} = 2R$) given by [21]

$$d_{eq} = \frac{4\sigma}{2\mu_0 \overline{M} H_I + \mu_0 M_I^2} \tag{5.8}$$

In these equations $\mathbf{M} \cdot \mathbf{H_I}$ is a scalar product.

Equation 5.8 shows the relative contribution of interfacial tension and magnetic properties respectively. For a fixed magnetic design, it is clear that d_{eq} can be reduced by increasing the magnetic field amplitude at the interface (increased when decreasing the distance between source magnets), increasing the magnetisation of the ferrofluid, or reducing the interfacial tension. Figure 5.3a summarizes how the antitube diameter varies with magnetic separation and the choice of ferrofluid [9]; with Fig. 5.3b we illustrate the experimental visualization by a bright field optical microscope of a small diameter antitube, using the magnet design of Fig. 5.1b and a water + surfactant antitube with MD4 ferrofluid and Tween20 surfactant. The annotation MD4 S denotes the ferrofluid MD4 and S the surfactant in the water anti-tube used to reduce the interfacial tension. The inset in Fig. 5.3b shows the gradual increase in ferrofluid volume to reach equilibrium diameter. Reduction in the equilibrium diameter was possible by using ferrofluid with higher saturation magnetization (EMG900) giving larger magnetic pressure. The ferrofluid EMG 900 (Ferrotec) is commercially available. Here EMG900 2S indicates EMG900 ferrofluid with double surfactant use in water and in ferrofluid [21]. A resulting $d_{eq} = 13 \pm 0.5$ μm was obtained, shown in Fig. 5.3c). Reduction of d_{eq}, can also be obtained by decreasing the size of the cell. Bringing the magnets closer (increasing M_I and $\overline{M}$) increases the magnetic pressure and further reduces the static diameter (d_{eq}). The experimental cell with the four-magnet arrangement resembles Fig. 5.1b with distance between magnet pair down to 150 μm. Figure 5.3d shows the experimental image of cross sections of the antitube (bright central part) and surrounding ferrofluid EMG S. EMG S denotes the ferrofluid EMG900 with surfactant (Tween20) in water to reduce the interfacial tension.

The minimum diameter of the antitube is 6.7 ± 0.2 μm, measured using synchrotron X-Ray tomography with 0.1 μm resolution. The antitube diameter data point is presented in Fig. 5.3a along with the data obtained by Dunne et al. [21] Our data is encircled (green) in Fig. 5.3a along with the data corresponding to Fig. 5.3b and c. Experiments agree well with the model curve [21]. LM and FM denotes the linear model and full model respectively. The linear model, LM ($M = \chi H$) is valid at small magnetic field whereas the full model considers the nonlinearity in the ferrofluid magnetization curve [21]. Note that antitubes larger than d_{eq} can always be made by injecting less ferrofluid. As such, d_{eq} is a lower limit of practical use for the antitubes.

In summary, the stability of such cylindrical antitubes is governed by the relative strengths of the destabilizing Laplace pressure and the stabilizing magnetic pressure. Since, the Laplace pressure is given by the ratio of interfacial tension and radius of the flow channel, it increases when the size of the fluidic circuit decreases. It therefore sets the limit for smallest achievable diameters when it exceeds the magnetic pressure. The latter depends on the magnitude of the applied local field and the magnetic susceptibility of the ferrofluid material. A suitable design of the magnetic field and the appropriate choice of intrinsic properties of the ferrofluid make possible the stabilization of minimal antitube size (d_{eq}). In a nutshell, one needs to maximize the magnetic susceptibility of the ferrofluid and minimize its interfacial surface tension with the enclosed liquid forming the antitube fluidic circuit. Our experiments show that diameters below 10 μm can be realized, with possible smaller values requiring further miniaturization of the source magnets (to increase the local magnetic field)

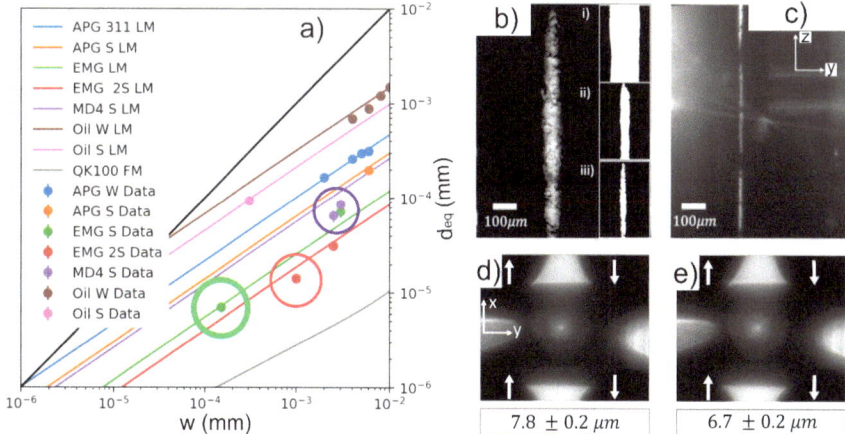

Fig. 5.3 Measurements of the static antitube minimum size. **a** Equilibrium diameter d_{eq} (Eq. 5.8) as a function of width W between the magnets, for several ferrofluids and antitube fluids (for abbreviation details, see [21]). **b** optical microscopy image of the antitube along the (yz) plane with MD4 ferrofluid. The inset shows the gradual filling of the cavity by the ferrofluid, reaching the data points shown in purple in **a** and **c** image of the antitube with EMG900 ferrofluid with added surfactant and water antitube with added surfactant in the double surfactant (EMG 2S) configuration giving the data point in red in (**a**). X-ray tomography circular cross sections shown in (**d**) and (**e**) are (xy) cut images of a water antitube in EMG900 ferrofluid with added surfactant, (single surfactant EMG S), with related green encircled data point indicating the smallest measured diameter of 6.7 μm in (**a**). Figure 5.3a is modified from the figure published in [21], with permission. Figure 5.3d and e were obtained using the X02DA TOMCAT X-ray beamline of Swiss Light Source (SLS) at the Paul Scherrer Institute, Villigen, Switzerland

and optimization of the ferrofluid intrinsic properties. Figure 5.3 indicates that the latter is possible, but imaging resolution issues make the observations of diameters below 1 μm elusive.

5.4   The Dynamic Case (Under Flow)

In the dynamic case, we detail the hydrodynamics resulting from the motion of the transported liquid. The conceptual difference between the standard flow (Poiseuille flow) and the proposed flow (antitube flow) is illustrated in Fig. 5.4.

A standard Poiseuille flow in Fig. 5.4 (top) is defined as a viscous flow with zero velocity at the confining wall (no-slip condition). This gives the well-known parabolic velocity profile with maximum velocity at the centre of the tube. In the proposed antitube flow with ferrofluid lubrication, the confining wall is a liquid (ferrofluid) shown in Fig. 5.4 (bottom). Since at the liquid–liquid interface, the velocity has to be unique (no slip condition), the confining liquid (ferrofluid) also moves along with the transported liquid. This makes lubricated flow, a system with moving wall. For small

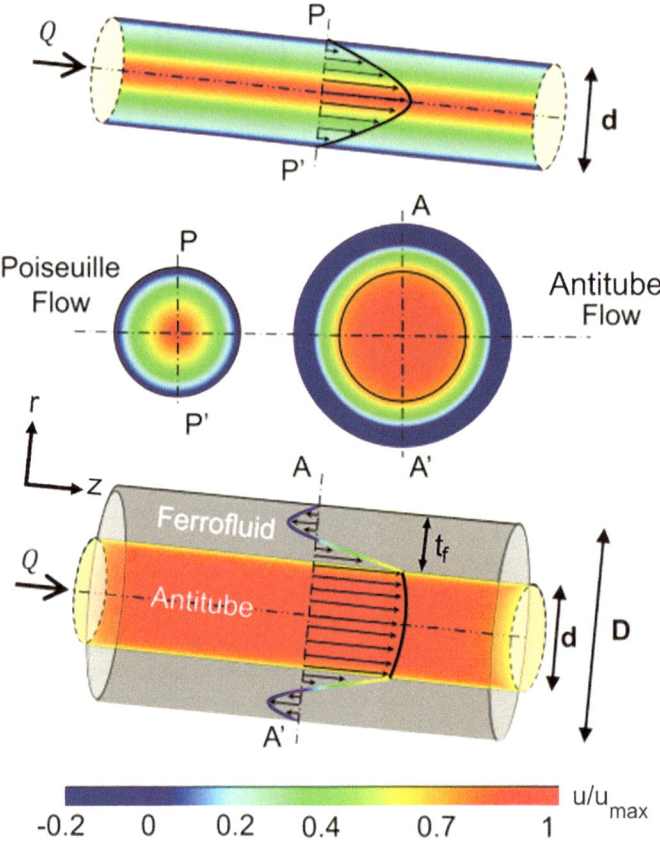

Fig. 5.4 Differences between Poiseuille flow and magnetically confined flow channels. Top: Poiseuille flow through a tube of diameter d, with flow rate Q, and the parabolic velocity profile showing zero velocity at the wall. Bottom: Magnetically confined flow channel (antitube) with transported liquid in antitube and ferrofluid encapsulation. The flow wall (between antitube and ferrofluid) has finite velocity. The ferrofluid exhibits a counter-flow circulation. D is the flow cavity diameter. Middle: Comparison of cross section of flow for the two flow systems; almost constant velocity (ideal flow) for magnetically confined flow compared to a uniform velocity gradient over the cross section for Poiseuille flow

enough flow rates, the magnetic force field holds the ferrofluid in place, avoiding its shearing out. The resulting volume conservation of ferrofluid implies a recirculation of the ferrofluid, depicted by the reverse flow in the darker region of the ferrofluid in Fig. 5.4. This proposed flow design results therefore in:

1. a finite velocity at the confining wall
2. a reverse flow path of the ferrofluid.

These two distinct features lead to spectacular high drag reduction values, measuring the reduction of friction forces, up to measured values of 99.8% for viscous liquids [20], or nearly frictionless flow. Figure 5.4 (middle) compares the flow velocity in the cross section for Poiseuille flow (left) and antitube flow (right). For the antitube, the velocity of flow is of nearly uniform amplitude across the flow channel, depicting near zero shear. The velocity profile (arrows) presented in Fig. 5.4 bottom is a result of solving the Navier–Stokes (N-S) equation with appropriate boundary condition, under the hypothesis of non-deformation of the cylindrical liquid-in-liquid flow, as observed experimentally for low enough flow rates.

In Fig. 5.4 (bottom), the interface between antitube (bright central part) and ferrofluid (darker encapsulation) is the liquid–liquid interface. The outer boundary of the ferrofluid is the solid boundary of the plastic cavity. A flow rate Q through a channel of width d is imposed for the transported liquid with the ferrofluid lubrication thickness being t_f. D is the width of the microchannel cavity. Since the Reynolds number is small (< 1), we can neglect the inertial terms in the N-S equation and the flow is viscosity-dominated, modelled and explained by the Stokes equation [20]. The wall velocity is given by [20]:

$$\frac{u_{wall}}{u_{max}} = \frac{\beta_0 - 1}{\beta_0 + 1} \tag{5.9}$$

$$\beta_0 = 1 + 4\eta_r \ln\left(1 + 0.8t_f^*\right) \tag{5.10}$$

where, β_0 is the simplified drag reduction factor [20] and η_r is the viscosity ratio of the two fluids. The normalized thickness of ferrofluid is described as $t_f^* = \frac{t_f}{d}$. The wall velocity depends on the relative coverage of the ferrofluid $\left(t_f^*\right)$ and the viscosity ratio (η_r). As expected, increasing the relative amount of ferrofluid improves the lubrication. Inversely, if the viscosity of the ferrofluid increases, the systems more and more resembles a solid wall boundary condition, therefore detrimental to lubrication.

Velocimetry experiments require no obstruction to the incident light (no ferrofluid in the path of light), hence a planar flow equivalent, slightly deviating from axisymmetric encapsulation of the transported liquid, is required. In this case, the ferrofluid covers the side of the flow channel, keeping the light unobstructed (see Fig. 5.5c).

We restrict ourselves to measure the flow profile away and equidistant from the two non-lubricated surface to minimize the influence of solid walls. Inset in Fig. 5.5a shows the schematic of the magnetic and flow arrangement, with a cross-section view of the four magnets in the quadrupolar arrangement and a top view showing the imposed flow rate Q along the length of the microchannel. The magnetic field generated by the arrangement is shown in Fig. 5.1b. The four white arrows in the cross section show the direction of magnetization of the magnets and the red line along the flow depicts the liquid–liquid interface. As required, the ferrofluid (colour grey) only covers the two walls and hence facilitates the transmission of light, essential for flow visualization.

We measure the velocity profile in the antitube directly using micro particle tracking velocimetry (μPTV). We use glycerol ($\eta_a = 1.1$ Pa.s) as antitube transported liquid and ferrofluids APGE32 ($\eta_a = 1.7$ Pa.s) from Ferrotec [22] and a biocompatible ferrofluid ($\eta_a = 0.144$ Pa.s) from Qfluidics [23] for testing the flow

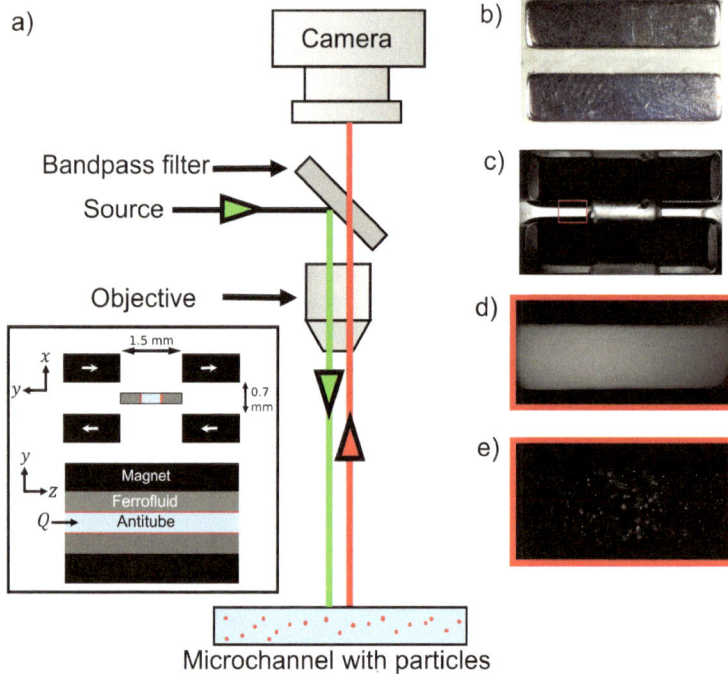

Fig. 5.5 Experimental set-up and direct measurement of velocity profile. **a** Optical system for Micro particle tracking velocimetry (μPTV) measurements, **b** microfluidic channel without ferrofluid, **c** microfluidic channel with ferrofluid, **d** brightfield image of region of interest under microscope, **e** fluorescence particles signal. Inset in Fig. 5.5a shows the schematic of microchannel and magnetic arrangement

behavior with different viscosity ratios, $\eta_r = \frac{\eta_a}{\eta_f}$. The antitube and the microchannel cavity width are $d = 0.5$ mm and $D = 1.5$ mm respectively. The ferrofluid coverage is maintained at, $t_f = \frac{D-d}{2}$. We begin by mixing glycerol with fluorescence 4 μ m size particles (FluoSpheres™). The excitation and emission maxima for the particles are 580 nm and 605 nm respectively. We use a source of light with a bandpass filter (572/25 for excitation and 629/62 for emission). Figure 5.5a shows the schematic of flow setup used for the velocimetry measurements. Figure 5.5b and c show the microfluidic channel before and after inserting the ferrofluid. Figure 5.5d shows a magnified bright field image of the region of interest for μPTV measurements and Fig. 5.5e shows particles under fluorescence in the micro channel.

The depth of field at this magnification is 3 μ m, below the chromophores diameter (4 μ m). This makes it possible to record images with minimum contribution from the off-focus particles (below or above the imaging plane). The time-elapsed image frames are recorded with Zeiss Axio zoom V16 microscope and a Phantom v2511 camera. The exposure time is 400 μs. The images are analysed by tracking differences in positions of particles in consecutive frames using FIJI TrackMate [24]. Data is then analysed in Python to bin the data for different heights and calculate the standard error for each bin height. Figure 5.6a) and b show the analytical Poiseuille flow and antitube experimental velocity profile using μPTV measurements. Markers are experimental and solid line stands for predicted analytical velocity profile [20]. Black lines denote the Poiseuille flow. The errors correspond to the standard deviation of velocities measured over 10,000 image sequences. The antitube lies between $y/d = [- 0.5\ 0.5]$. The Poiseuille flow shows the zero velocity at the wall coordinates, $y/d = 0.5$ and $y/d = - 0.5$. Experimental markers in Fig. 5.6a show that for $\eta_r = 0.65$ with APGE32 as ferrofluid, the wall velocity at the liquid–liquid interface is large enough, up to 60% of the maximum velocity. This wall velocity reaches almost 85% of the maximum velocity for $\eta_r = 7.64$ as seen in Fig. 5.6b for a biocompatible ferrofluid. The ferrofluid coverage is slightly asymmetric for $\eta_r = 0.65$, as the ferrofluid thickness are a bit different between left and right. For $\eta_r = 7.64$, the ferrofluid coverage is symmetric. The increase of wall velocity with increase in η_r is also forecast by the analytical predictions.

Figure 5.6a and b show direct evidence of large wall velocity with a magnetic fluid lubrication as previously predicted by several of us [20, 21]. The increase of wall velocity with increase in η_r is also foreseen by the analytical predictions and relates to the expected enhanced lubrication. The wall velocity values predicted for APGE32 (Fig. 5.6a) and the biocompatible ferrofluids (Fig. 5.6b) given by Eq. 5.8 are 32% and 90% of the centre velocity maximum respectively.

The experimental and analytical predictions agree with some deviations. The experimental measurements might differ from that theorized in literature [20] due to presence of two solid walls, which is not taken into account in the analytical modelling [21]. Nonetheless, the direct measurements do confirm the presence of large wall velocity at the liquid–liquid interface which were only indirectly claimed earlier [20, 21]. We have limited ourselves here to the regime of low flow rate, to avoid capillary instabilities that can develop at higher flow rates. This may result

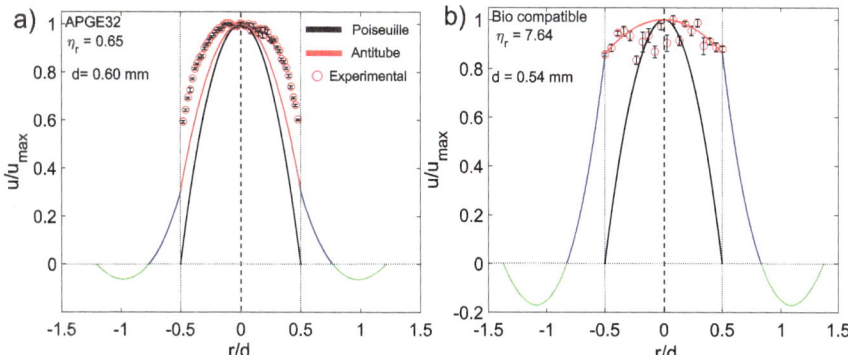

Fig. 5.6 Comparison of experimental velocity profiles (markers) with Poiseuille flow (Black line). **a** For APGE32 ferrofluid with viscosity ratio $\eta_r = 0.65$ and **b** for biocompatible ferrofluid with $\eta_r = 7.64$. Lines are the analytical model solution of Stokes equation, red line for the antitube, green and blue line for the ferrofluid

in deformations in the magnetic-nonmagnetic interface, large enough to result in shearing of the ferrofluid or shear failure of the lubricated flows. A more sophisticated analytical approach and numerical simulations are necessary to extend this simple model to include liquid wall deformation or to understand flow behaviour at the extremities of the circuit, beyond the scope of the present work. Instead, we focus on testing how solid cavity design changes can impact the antitube flow profile, with an example discussed in the next section.

5.5   Beyond a Simple Cylindrical Symmetry

We argue that the symmetry of the flow can be broken by implementing deformations all along the chamber walls (i.e., only in contact with the magnetic fluid). One possible implementation is to define a threaded screw profile along the cylindrical circuit length (Fig. 5.7a), where a small rectangular section was left free, to allow for proper imaging using confocal microscopy. The magnetic design is given in the inset of Fig. 5.7a. Figure 5.7b illustrates the occurrence of a vortex flow profile resulting from the screw-shaped solid walls boundary conditions. The prediction of fluid motion with complex fluid interfaces poses a challenge. Here, we qualitatively explain our experimentally observed flow in a screw wall device using a forced vortex model. A forced vortex in two dimensions (2D) implies the occurrence of an azimuthal velocity V_θ^* simply expressed as [25]

$$V_\theta^* = \omega^* r^* \qquad (5.11)$$

It is proportional to the radial position $r^* = \sqrt{x^2 + y^2}$ through the angular velocity ω^*, with all variables non-dimensional and reduced by their values at the magnetic-nonmagnetic interface (flow wall).

Figure 5.7c shows the azimuthal velocity with respect to the radial coordinate. It can be seen from Fig. 5.7c that the experimental data obtained for the azimuthal velocity matches well with the theoretically predicted fit for $r^* > 0.5$, with flow similar to a forced vortex. The inset in Fig. 5.7c shows that the axial velocity (V_z) (in the direction of flow) decreases sharply after $r^* > 0.45$. This reduced axial velocity makes a 2D model, implicit in the Eq. 5.10, relevant. Indeed, a key signature of a forced 2D vortex flow is the constant vorticity given by, $\Gamma^* = 2\omega^*$. Figure 5.7d shows the vorticity plot, illustrating the transition from a 3D to a 2D behaviour when r^* reaches 0.5. The blue highlighted region in Fig. 5.7c and in the inset show the change in behaviour of azimuthal velocity and axial velocity respectively with r^*, confirming the change of behaviour at $r^* \approx 0.5$ highlighted in Fig. 5.7d. We therefore find that the flow has two distinct behaviours; the flow is 2D-dominated when $r^* > 0.5$ (lower axial velocity) and possibly 3D-augmented for $r^* < 0.5$.

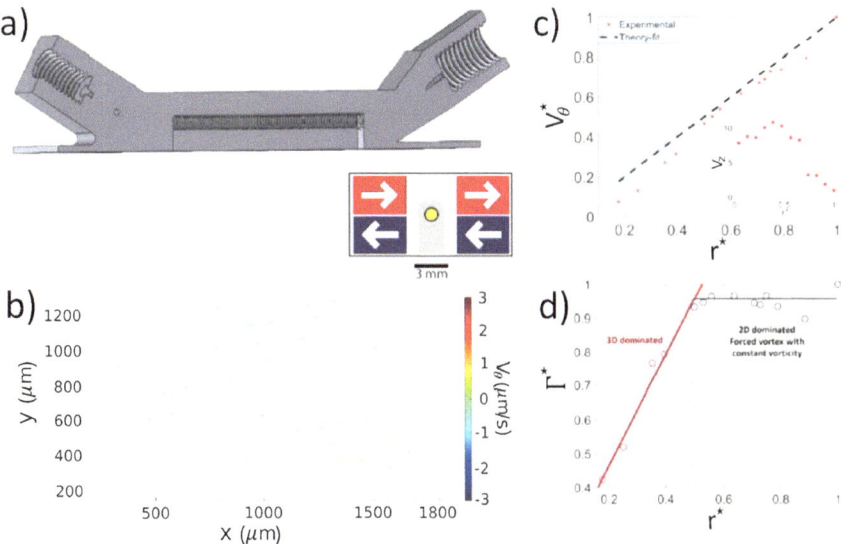

Fig. 5.7 Vortex flow in antitube. **a** Screw walled device for vortex flow design, inset shows configuration of four magnets in the design with yellow central part depicting the antitube, **b** vortex
flow observed by velocimetry, **c** forced vortex model and experimental data, inset shows the axial
velocity (V_z) in the direction of flow with respect to the radial coordinate and **d** experimental
vorticity $\Gamma^*(= 2\omega^*)$ as a function of reduced radial length

5.6   Conclusions

Ferrofluid encapsulation of liquid circuits opens new possibilities for microfluidics
and fluidic applications [21]. Stabilizing an encapsulated cylindrical 'antitube' of
dimensions below 10 μm is achieved. This size value is governed by the liquid–
liquid interface surface tension, as well as the scale of the surrounding assembled
magnetic force field sources and the magnetic properties of the ferrofluid. Values
below 1 μm are possible, but very challenging to image. Magnetic forces keeping
the lubricating magnetic liquid in place mitigate the fundamental issue of stability of
a liquid-in-liquid flow [15]. This makes possible to extend achievable drag reduction,
resulting in remarkably large values, possibly exceeding 99% [20]. A direct measurement of velocity profile using μ PIV confirms the occurrence of liquid–liquid wall
velocities as large as 85% of the maximum velocity. These values increase with
the increase in viscosity ratio between the transported liquid to the lubricant. The
experimental observations are supported and predicted by analytical modelling using
the Stokes equation. Such low drag flow channels enable microfluidic applications
that require reduced operating pressures, specifically needed in microchannel flow of
concentrated solution or variable viscosity solutions. Drug delivery and shear control
on delicate cells also falls in the gamut of possible applications. The magnetically
confined flow channels therefore open new possibilities in the field of microfluidics

for shearless transport in a wide range of biological and technological applications. Beyond a simple Poiseuille-like parabolic flow profile, complex flow profiles can be obtained by tuning the design of the cavity used for the flow. We have shown that a screw pattern on the inner boundary of the cavity results in swirling motion of flow about the central axis. The flow resembles a forced vortex flow with constant vorticity and deviates as we approach the centre of the flow, away from the screw design. These examples provide insight into the possibility to design and investigate novel flow patterns, otherwise not possible when using the solid walls that constrain the flow of the transported liquid.

Acknowledgements This project has received funding from the European Union's Horizon 2020 research and innovation programme under the Marie Skłodowska-Curie grant agreement MAMI No 766007 and QUSTEC No. 847471. We also acknowledge the support of the University of Strasbourg Institute for Advanced Studies (USIAS) and the Fondation Jean-Marie Lehn, as well as the support from IdEx Unistra (ANR 10 IDEX 0002), SFRI STRAT'US project (ANR 20 SFRI 0012) and EUR QMAT ANR-17-EURE-0024 under the framework of the French Investments for the Future Program. We acknowledge the Paul Scherrer Institute, Villigen, Switzerland for synchrotron radiation beam time at the TOMCAT beamline X02DA of the SLS and would like to thank Dr. Anne Bonnin for invaluable assistance.

References

1. E. Karatay, A. Haase, C.W. Visser, C. Sun, D. Lohse, P. Tsai, R. Lammertink, Control of slippage with tunable bubble mattresses. Proc. Natl. Acad. Sci. U. S. A. **110** (2013)
2. W. Brostow, Drag reduction in flow: Review of applications, mechanism and prediction. J. Ind. Eng. Chem. **14**, 409 (2008)
3. K. Watanabe, Y. Udagawa, H. Udagawa, Drag reduction of Newtonian fluid in a circular pipe with a highly water-repellent wall. J. Fluid Mech. **381**, 225 (1999)
4. H. Hu et al., Significant and stable drag reduction with air rings confined by alternated superhydrophobic and hydrophilic strips. Sci. Adv. **3**, e1603288 (2017)
5. D. Saranadhi, D. Chen, J.A. Kleingartner, S. Srinivasan, R.E. Cohen, G.H. McKinley, Sustained drag reduction in a turbulent flow using a low-temperature Leidenfrost surface. Sci. Adv. **2**, e1600686 (2016)
6. K. Holmberg, A. Erdemir, Influence of tribology on global energy consumption, costs and emissions. Friction **5**, 263 (2017)
7. K. Sayfidinov, S.D. Cezan, B. Baytekin, H.T. Baytekin, Minimizing friction, wear, and energy losses by eliminating contact charging. Sci. Adv. **4**, eaau3808 (2018)
8. J. Keeling, N.G. Berloff, Going with the flow. Nature **457**, 7227 (2009)
9. A. Keerthi, S. Goutham, Y. You, P. Iamprasertkun, R.A.W. Dryfe, A.K. Geim, B. Radha, Water friction in nanofluidic channels made from two-dimensional crystals. Nat. Commun. **12**, 1 (2021)
10. H.M. López, J. Gachelin, C. Douarche, H. Auradou, E. Clément, Turning bacteria suspensions into superfluids. Phys. Rev. Lett. **115**, 028301 (2015)
11. Jayaprakash, Enhancing the injectability of high concentration drug formulations using core annular flows. Adv. Healthc. Mater. Wiley Online Library (2020). https://onlinelibrary.wiley.com/doi/abs/10.1002/adhm.202001022
12. T.-S. Wong, S.H. Kang, S.K.Y. Tang, E.J. Smythe, B.D. Hatton, A. Grinthal, J. Aizenberg, Bioinspired self-repairing slippery surfaces with pressure-stable Omniphobicity. Nature **477**, 7365 (2011)

13. B.R. Solomon, K.S. Khalil, K.K. Varanasi, Drag reduction using lubricant-impregnated surfaces in viscous laminar flow. Langmuir **30**, 10970 (2014)
14. L. Rayleigh, On the stability, or instability, of certain fluid motions. Proc. Lond. Math. Soc. **s1–11**, 57 (1879)
15. A.S. Utada, A. Fernandez-Nieves, H.A. Stone, D.A. Weitz, Dripping to jetting transitions in coflowing liquid streams. Phys. Rev. Lett. **99**, 094502 (2007)
16. J.S. Wexler, I. Jacobi, H.A. Stone, Shear-driven failure of liquid-infused surfaces. Phys. Rev. Lett. **114**, 168301 (2015)
17. M.S. Krakov, E.S. Maskalik, V.F. Medvedev, Hydrodynamic resistance of pipelines with a magnetic fluid coating. Fluid Dyn. **24**, 715 (1989)
18. V.F. Medvedev, M.S. Krakov, E.S. Mascalik, I.V. Nikiforov, Reducing resistance by means of magnetic fluid. J. Magn. Magn. Mater. **65**, 339 (1987)
19. A. Blaeser, D.F. Duarte Campos, U. Puster, W. Richtering, M.M. Stevens, H. Fischer, Controlling shear stress in 3D bioprinting is a key factor to balance printing resolution and stem cell integrity. Adv. Healthc. Mater. **5**, 326 (2016)
20. A.A. Dev, P. Dunne, T.M. Hermans, B. Doudin, Fluid drag reduction by magnetic confinement. Langmuir **38**, 719 (2022)
21. P. Dunne et al., Liquid flow and control without solid walls. Nature **581**, 7806 (2020)
22. Ferrotec|Manufacturing advanced material, component, and system solutions for precision processes. https://www.ferrotec.com/
23. Qfluidics—Continuous flow chemistry reactor manufacturer. https://www.qfluidics.com/
24. D. Ershov et al., TrackMate 7: integrating state-of-the-art segmentation algorithms into tracking pipelines. Nat. Methods **19**, 7 (2022)
25. S.I. Green, *Fluid vortices*, 1st edn. (Springer, Dordrecht, 1995)

Chapter 6
Hematite Cubes

M. Brics, O. Petrichenko, and Andrejs Cēbers

6.1 Introduction

Iron(III) oxide Fe_2O_3 at ambient conditions has four crystalline polymorphic forms: α-Fe_2O_3 (hematite), β-Fe_2O_3, γ-Fe_2O_3 (maghemite), and ϵ-Fe_2O_3. These forms have distinctly different structural and magnetic properties. The most common of them is hematite.

Hematite has a rhombohedral crystal structure isostructural with corundum (α-Al_2O_3). Below the Morin temperature [1], for bulk hematite $T_M \approx 260 \, k$ [2], hematite is an antiferromagnetic material. Spins reorientate under increasing temperature and, due to the Dzyaloshinsky-Moriya mechanism [3, 4], hematite becomes a weak ferromagnetic material. It remains a weak ferromagnetic material up to the Néel temperature, which for bulk hematite is $T_N \approx 950 \, K$ [5].

Another interesting property of hematite is that hematite particles maintain a permanent dipole moment even at large sizes (up to 15 μm) [6, 7]. Thus, it allows at room temperature to create magnetic colloids (made of single domain magnetic particles) which can be directly observed with an optical microscope [8–12]. These colloids allow to investigate an interesting physical regime where magnetic forces, hydrodynamic forces, steric forces, and thermal fluctuations are comparable quantities and thus important to describe dynamics. Moreover, the colloidal particles can be synthesized in different shapes: cubes, disks, ellipsoids, peanuts, needles, …[8, 13–15].

In this paper we mainly focus our discussion on colloids made of hematite cubes, as cubic-shaped hematite particles have an unorthodox magnetization orientation.

Supplementary Information The online version contains supplementary material available at https://doi.org/10.1007/978-3-031-58376-6_6.

M. Brics · O. Petrichenko · A. Cēbers (✉)
MMML Lab, Department of Physics, University of Latvia, Jelgavas 3, Rīga, LV-1004, Latvia
e-mail: andrejs.cebers@lu.lv

© The Author(s) 2024
B. Doudin et al. (eds.), *Magnetic Microhydrodynamics*, Topics in Applied Physics 120,
https://doi.org/10.1007/978-3-031-58376-6_6

Fig. 6.1 The magnetic moment orientation in a hematite cube. The angle $\phi = 12°$ is in the plane defined by two diagonals and the magnetic moment $\boldsymbol{\mu}$ points to the face

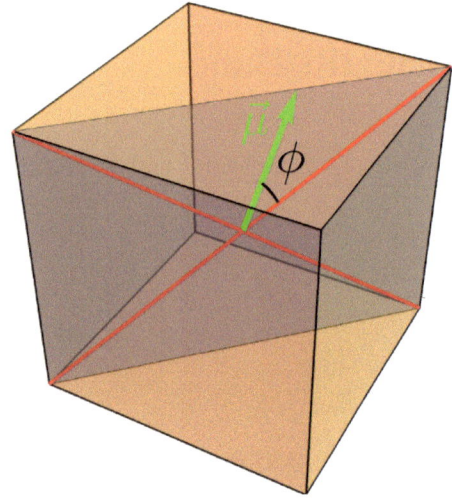

The magnetic moment with a cube's diagonal makes an angle 12° (see Fig. 6.1) in the plane defined by two diagonals [6, 11].

In the scientific literature several very interesting experiments with hematite colloids formed by cubic particles can be found. Recently Chen et al. [16] investigated the medical application of cubic-shaped hematite microrobots. The main goal was the sweep of microblocks and impurities in blood vessels. There authors demonstrated that approximately $2\,\mu m$ large cube-shaped hematite particles can be guided by a rotating magnetic field. In the xz (vertical) plane rotating magnetic field introduced a rolling motion and the cube moved along the x axis. In such a way cubes were able to overcome obstacles and push small objects. Similarly the motile structures formed by microrollers which were created by micron sized polymer colloids with embedded hematite cubes were demonstrated in [17].

Soni et al. [18] showed that a two-dimensional chiral fluid can be created using hematite colloids. The densely packed ensemble of hematite cubes in a horizontal plane rotating magnetic field behave like a two-dimensional fluid showing characteristic instabilities. In the article [19], authors demonstrated targeted assembly and synchronization of self-spinning microgears or rotors made of hematite cubes and chemically inert polymer beads. In [20], a potential application of hematite colloidal cubes for the enhanced degradation of organic dyes was investigated. In [21], the formations of light activated two-dimensional "living crystals" was examined.

In this article we summarize the synthesis processes of hematite cubes and results of experiments on particle structures in hematite colloids based on our experiments [6, 12, 22, 23] and the work done by the group of Albert P. Philipse from Utrecht University and their collaborators [8, 11, 15, 18, 24–26].

The content of this paper is divided into four sections. The Sect. 6.1 is an introduction followed by a Sect. 6.2 where synthesis methods for hematite cube are described.

The particle structures in hematite colloids are summarized in Sect.6.3 and conclusions in Sect. 6.4.

6.2 Synthesis

The fabrication of the hematite, α-Fe_2O_3 nanosized particles with desired morphologies attracts attention due to their interesting optical, chemical and magnetic properties [27, 28]. A large number of micrometer particles of metal oxides have been developed as models for research colloid science and advanced materials [29].

Böhm in 1925 first found that freshly precipitated amorphous Fe^{3+} hydroxide turns into goethite if kept for 2 h under 2M KOH at 150 °C, whereas hematite is the dominant end-product if material is heated in the water [30].

Matijević and co-workers [31] have developed and described in detail the preparation of ferric hydrous oxide sols consisting of colloidal hematite particles uniform in shape. They reported the preparation of cubic, ellipsoidal, pyramidal, rod-like, and spherical hematite particles. Matijević and Scheiner demonstrated that minor changes in the reaction environment could produce significant changes in the morphology of iron oxide particles. The importance of this work was due to the fact that for the first time well-defined monodisperse colloids of general metal hydrous oxides were developed and described. However, hematite particles were obtained from dilute homogeneous solutions of concentrations of concentration of the order of 10^{-2} M or less.

The gel-sol method proposed by Sugimoto has shown the possibility to prepare monodispersed hematite particles precisely controlled in shape and size, from highly condensed ferric hydroxide gel. The advantage of this method compared to the dilution solution method is the high productivity and the high yield of hematite without any solid byproduct such as β-FeOOH [29, 32].

Commonly, α-Fe_2O_3 particles can be prepared by controlled hydrolysis of ferric salts and hydroxides carried out via solvothermal/hydrothermal techniques:

- forced hydrolysis of diluted $FeCl_3$ aqueous solutions which are kept for a certain time (from 3 to 192 h) [27, 33] at elevated temperatures.
- a diluted solution method applying low concentration solutions of ferric salts and alkali hydroxides used as initial reactants [31];
- a gel-sol method which allows preparing monodispersed hematite particles precisely controlled in shape and size from highly condensed ferric hydroxide gel [32, 34, 35].

The formation of hematite proceeds by transformations of iron hydroxide ($Fe(OH)_3$ through akaganeite (β-FeOOH) to hematite (α-Fe_2O_3) [33, 34].

Solvothermal/hydrothermal techniques are popular methods for hematite particles production with varying size and morphology, by changing the growth parameters

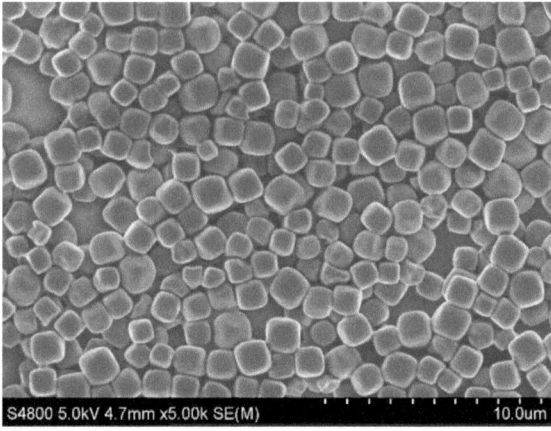

Fig. 6.2 Images of cubic-shaped hematite particles obtained by SEM. The mean particle size is $1.28 \pm 0.38\,\mu$m. The scale bar is $10\,\mu$m

such as different solvent combination for precursor solution [28, 36], and by varying the excess concentration of Fe^{3+} ions relative to that of hydroxide $(OH)^-$ ions. Their size is increasing with Fe^{3+} ions concentration [31, 37], which is in agreement with [33], where the particles growth in ferric chloride solutions was investigated under forced hydrolysis conditions. It was found that, depending upon the initial $FeCl_3$ concentration, either small single crystal hematite nanocubes or larger pseudocubic polycrystalline hematite particles form.

Park et al. [29] investigated morphology and internal structure of monodispersed pseudocubic hematite particles produced by the gel-sol method through high-resolution electron microscopy. It was found that sub-crystals of cubic-shaped hematite particles are radially developed from the center of a particle in all directions, but most preferentially in the directions of the longest diagonal axis of a particle. The longest diagonal of a pseudocubic particle corresponds to the c-axis.

The α-Fe_2O_3 micro-sized particles presented in the Figs. 6.2 and 6.3 were prepared via the standard gel-sol method of Sugimoto et al. [32, 34] with template method small adjustment [8] by the following procedure:

- A sodium hydroxide aqueous solution (21.64 g NaOH/ 100 ml H_2O) at rate of 5 ml/min was gradually added into an iron chloride hexahydrate aqueous solution (54 g $FeCl_3 \cdot 6H_2O$/100 ml H_2O). This solution was under vigorous magnetic stirring. The resulting dark brown gel was stirred additionally for 5 min.
- Obtained precursor was hermetically closed in a Pyrex bottle and placed into a laboratory oven at 100 °C for aging and left undisturbed for 7 d.
- The resulting precipitated solids were washed by distilled water through centrifugal separation and ultrasonic re-dispersion in water until reddish-brown color hematite particles were obtained.
- The hematite particles were dispersed in the distilled water, stabilized with sodium dodecylsulfate (Na$C_{12}H_{25}SO_4$, SDS) (0.11 g SDS/80 ml H_2O) and finally adjusted by tetramethylammonium hydroxide ((CH_3)$_4$NOH, TMAOH) aqueous solution

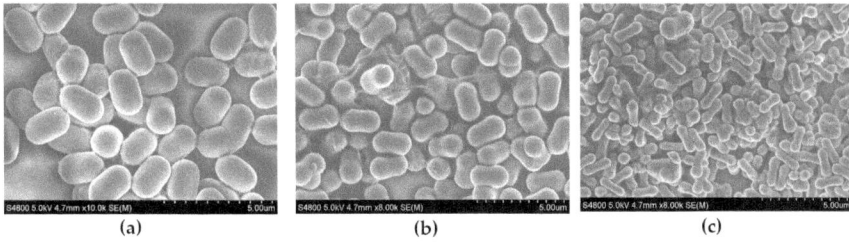

Fig. 6.3 SEM images of α-Fe$_2$O$_3$ particles of different shape prepared under the standard Sugimoto' gel-sol route for cubic-shaped particles but with addition of SO$_4^{2-}$ ions to the precursor mixture (see Table 6.1). **a** ellipsoids, the mean particles size is $2.10 \times 1.34 \, \mu$m. **b** big ellipsoids, the mean particles size is $2.23 \times 1.15 \, \mu$m. **c** "peanuts", the mean particles size is $1.68 \times 0.48 \, \mu$m. The scale bar is $5 \, \mu$m

Table 6.1 Precursors composition for different shaped hematite particles (see Figs. 6.2 and 6.3) prepared by the gel-sol method

particle' shape	NaOH/100 ml H$_2$O(g)	Na$_2$SO$_4$/10 ml H$_2$O (g)
Pseudocubes	21.64	–
Ellipsoids	21.64	0.285
Big ellipsoids	19.48	0.285
"Peanuts"	19.48	0.869

The amount of FeCl$_3 \cdot$ 6H$_2$O in 100 ml H$_2$O is 54 g [8, 32, 34]

to 8.5–9.5 values. This procedure prevents hematite particles from irreversible sticking onto the glass surface induced by attractive Van der Vaalse interactions [12, 38].

The preparation of ellipsoid and peanuts-shape hematite particles shown in Fig. 6.3 was similar to the procedure described above for cubes except that for NaOH and Na$_2$SO$_4$ amounts. Conditions for preparation of different shape hematite particles are presented in the Table 6.1.

The obtained hematite particles were characterized using scanning electron microscopy (SEM) Hitachi S4800 to investigate their size. Magnetic properties for dried cubic-shaped hematite particles were determined by a vibrating sample magnetometer Lake Shore Cryotronics, Inc. 7400 VSM. The magnetization curve shown in Fig. 6.4 indicates a clear hysteresis and notable coercivity $B_c = 250$ mT. Remanent magnetization is around $M = 1.9$ kA/m.

In [34, 37] it was reported that the morphology of cubic-shaped hematite particles of the order of 1 μm in size obtained by aging a condensed ferric hydroxide gel at 100 °C for 7–8 d (see Fig. 6.2) can be modified from cubic via ellipsoidal to peanut-shape by introducing increasing amounts sulfate ions into the ferric hydroxide gel.

It was found that sulfate ions restrain the growth in all directions normal to the c-axis. The anisotropy was explained in terms of specific adsorption of sulfate ions onto the sub-crystals of each hematite particle, retarding the surface reaction of ferric complexes such as Fe(OH)$^{2+}$ on the planes perpendicular to their c-axis [29, 34].

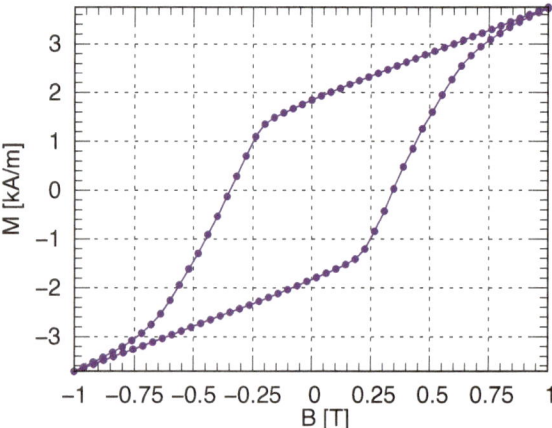

Fig. 6.4 Cubic-shaped hematite sample magnetization curve

6.3 Results

In colloids, hematite has a density $\rho_h = 5.25$ g/cm^3 significantly larger than the solvent (usually water $\rho_s = 1.00$ g/cm^3). Thus, micron-sized hematite particles sediment [6]. If there is no external magnetic field after the sedimentation process, hematite cubes lie on a face. However, under horizontal magnetic field $B > 15\,\mu$T after the sedimentation process, cubes stand on their edges [22]. For an individual cube there are two alignments (see Fig. 6.5) how it can lie [6]. The second alignment can be obtained from the first one by rotating the cube by 180° around an axis parallel to the magnetic field (x axis in Fig. 6.5).

In a weekly concentrated colloid under a static horizontal external magnetic field, the hematite particles form chains [6, 8, 11]. Depending on the strength of the

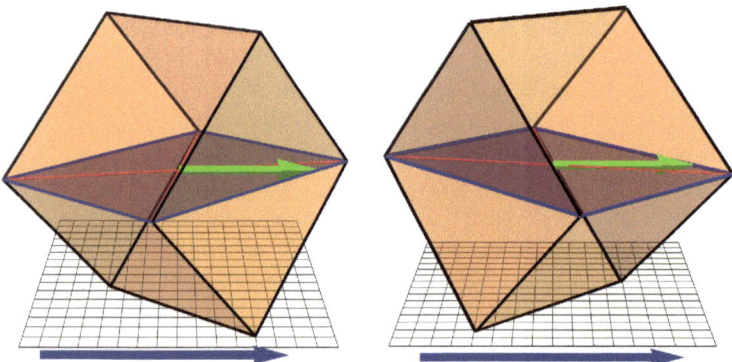

Fig. 6.5 Two alignments how the cube can lie on a surface after sedimentation. The green and blue arrow shows the orientation the magnetic moment and the direction of the magnetic field respectively

Fig. 6.6 Chain with kinks (top) versus straight chain above B_c (middle) versus straight chain below B_c (bottom). The top chain has two kinks. The blue arrow shows magnetic filed direction and little green arrows show magnetic moment orientation (zig-zag structure) in the bottom chain. For all other chains in this figure magnetic moments are along magnetic field

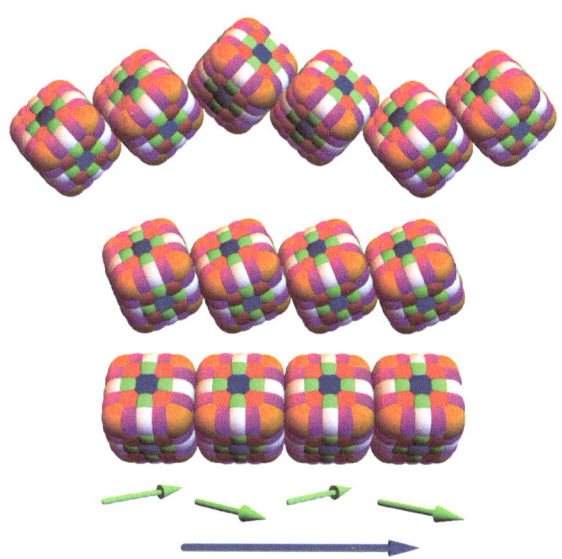

external magnetic field, two chain configurations are observed. Below the critical magnetic field $B_c \approx 0.1$ mT [6], which is typically larger but comparable with the Earth magnetic field, the first configuration is observed. In this case cubes arrange in straight chains (Fig. 6.6) and magnetic moments form zig-zag structures. The second configuration is observable for $B > B_c$ and the magnetic moment of every cube is parallel to the external magnetic field. The short chains are predominantly straight, however, longer chains contain kinks (Fig. 6.6). The kinks are formed during the assembly process when two cubes (two chains, a cube and a chain) with different alignments attach. For the short chains no kinks are experimentally observed as the thermal energy is sufficient for rotation of a single cube (changing cubes alignment) and a straight chain is formed. However, the thermal energy is not sufficient to change alignment of two or more cube chain, thus longer chains contain kinks [6].

Therefore, in a static external magnetic field, three chain types are observed which are shown in Fig. 6.6. For magnetic fields below B_c straight chains with magnetic moments arranged in a zig-zag structure are observed. Due to thermal effects the chain fluctuates (more pronounced for shorter chains), but the average angle between chain direction and the external magnetic field $\theta = 0°$ [6]. For magnetic fields above B_c chains with magnetic moments parallel to the external field are found. Short chains (less than five cubes) are usually straight and longer chains have kinks. For straight chains cubes are shifted against each other. These chains also fluctuate due to thermal effects, however, fluctuations are less pronounced as the magnetic field strength is larger. Also the average angle θ changes and reaches $\theta \approx \pm 18°$ [6]. For the chain with kinks the angle θ depends on the arrangement of the chain. However, for the straight parts of the chain one finds that average $\theta_i \approx 16.3°$ [11], which is close to the short chain value [6].

If a rotating magnetic field is applied, individual cubes and chains rotate [22] or roll [16]. The rotation and rolling motion are observed if the magnetic field is applied in the xy (horizontal) or xz (vertical) plane respectively. If the magnetic field is applied in the xz plane, then rolling motion of the first chain configuration (Fig. 6.6) is observed. If the rotating magnetic field is applied xy plane, then all three types of chains can rotate, however, kinked chains are very fragile and easily break and straight chains with magnetic moments orientated in zig-zag structure are stable only for small magnetic fields. Thus, in experiments mostly short chains with aligned magnetic moments are found [22].

Two scenarios of cube rotation depending on the frequency of the rotating magnetic field are possible. For a frequency smaller than the critical frequency f_c, the cube rotates synchronously with the magnetic field. Depending on the strength of the magnetic field and initial conditions the cube with rounded corners rotates on an edge, corner, or a face [22]. The corresponding motion of the cube can be seen in Video1, Video2, and Video3 [39]. For magnetic field frequency $f > f_c$, an asynchronous motion of a cube is observed. The critical frequency $f_c \propto 1/B$ and for 1.5 μm large hematite cubes at $B = 1\,\text{mT}$ one finds that $f_c \approx 10\,\text{Hz}$.

For large enough rotating magnetic fields, depending on the initial conditions, either precession of the magnetic moment or back-and-forth rotation are observed (see Video4 and Video5 [39]). In the last case, the cube rotates more slowly than the magnetic field and, in order to catch up with the magnetic field, the cube for a short time rotates in the opposite direction. When gravitational effects start to dominate, precession is not observed any more. Instead, a combination of back-and-forth and precession is observed (see Video6 [39]). Initially a cube rotates on its face. The lag increases, but instead of back motion to catch up with the magnetic field, the cube rolls, the magnetic moment goes out of the plane of the rotation magnetic field and through this rolling motion catches up with the magnetic field [22]. For a single cube, the magnetic moment usually goes out of the plane of the rotation magnetic field. The magnetic moment is in the plane of the magnetic field only for synchronous rotation on an edge (see Video1 [39]), where the edge slides on the bottom surface of a capillary [22].

For chains of cubes, depending on the frequency of the rotating magnetic field, the two scenarios of synchronous and asynchronous rotation with the magnetic field are possible. The critical frequency f_c depends on the chain length (number of cubes in chain), decreasing for increasing chain length. In the synchronous motion, the cubes forming chains rotate on an edge or a face (see Video7 and Video8 [39]). No motion on a corner is possible due to geometric restriction. In the asynchronous case back-and-forth rotation, periodic disassembly and reassembly of chain, and out of plane rotation are observed [22]. The corresponding motion of the two-cube chain can be seen in Video9, Video10, and Video11 [39]. The precession of the magnetic moment for a chain is not observed due to the same geometric restrictions. Instead, the out of plane rotation was observed. For out of plane motion, one finds the similar dynamics as for a single cube when the precession of the magnetic moment become impossible. To catch up with the magnetic field, the chain rolls and catches up with the magnetic field through rolling motion where the magnetic moment goes out of the plane of the rotating magnetic field and then returns. Unlike, for a single cube, magnetic moments of cubes in a chain are usually in the plane of the rotation magnetic field. The magnetic moment goes out of the plane of the rotating magnetic field only in one case, called the out of plane rotation, which is a combination of precession and back-and-forth motion. [22].

Interestingly, the dynamics of an individual chains depend on the clockwise or anticlockwise rotation direction of the magnetic field [22]. However, there is no more trend when averaging over many chains. The reason for this is that, similarly to single cube (see Fig. 6.5), also two alignments are possible for straight chains in a static magnetic field. In a large sample there are approximately equal number of chains in each alignment. Particles in each alignment behave differently at a given clockwise and anticlockwise rotation direction of the magnetic field. But the first alignment's dynamics in a clockwise rotating magnetic field is equal to the second alignment's dynamics in an anticlockwise rotating magnetic field. Thus, they balance out this effect and on average there are no differences [22].

Hematite particles agglomerate when their concentration is increased in a static magnetic field [8, 12]. In a slowly rotating magnetic field (in xy plane) agglomerates rotate as a solid body with the frequency of the rotating magnetic field. If the frequency is increased $f \in (3, 30)$ Hz the swarms are formed [12, 18]. Swarms of circular shape consist of individual rotating cubes and short chains (mostly two-cube and three-cube chains) [12]. Chains and individual cubes forming swarms rotate with the frequency of the rotating magnetic field [18]. Swarms exhibit a smaller rotation frequency than the rotating magnetic field. For large swarms (consisting off more than 10^5 cubes) only the outer particles rotate as the rotation speed is an exponential function from the distance to the center of the swarm [18]. For smaller swarms (10^3–10^4), however, particles in the center rotate with almost angular constant velocity which agrees with the results of model [23], which incorporates lubrication forces and magnetic dipole-dipole interactions. For small frequencies (up to 15 Hz) the rotation frequency of the swarm is proportional to the frequency of the rotating magnetic field [12, 18]. The large swarms behave like a two-dimensional chiral fluid [18].

6.4   Conclusions

After summarizing the literature of hematite synthesis methods, particular attention was paid to the methods for the production of cubic-shaped hematite particles. Solvothermal methods in particular the diluted solution method and Sugimoto gel-sol method were identified as the most suitable ones. It has been proved that the size of hematite particles strongly depends on the Fe^{3+} ions concentration in the initial solution.

We also summarized the richness of the possible structures that cubic-shaped hematite particles may form in magnetic colloids. In weakly concentrated colloids, particles arrange in chains of types that depend on the strength of the external magnetic field: straight chains with magnetic moments arranged in zig-zag structures, short straight chains with magnetic moments aligned with the external magnetic field, and longer chains with kinks, while keeping the individual particles magnetic moments aligned with the external magnetic field. In a rotating magnetic field chains and individual cubes rotate or roll. Rolling motion and rotation are observed when the magnetic field is rotating in the xz (vertical) and xy (horizontal) plane respectively. Both for rolling and rotation synchronous and asynchronous motion with the external rotating field is observed. In the case of rotation, the chain can go out of the plane of the rotating magnetic field.

For magnetic colloids with higher concentration, particles arrange and form aggregates. These aggregates rotate as a solid-body in a very slowly rotating magnetic field. If the frequency is increased the circular shaped swarms of particles are formed. These swarms consist of individual cubes and short chains which rotate with the frequency of the external magnetic field.

Suspension of hematite particles is interesting, in particular, by unusual competition of magnetic and steric interactions due to the non-trivial orientation of the magnetic moment in the particle. The size of the hematite particles makes it possible to observe them easily in the optical microscope. Since their magnetization is rather small then new situations arise where magnetic interactions compete with other interactions in the system, for example, due to the viscous lubrication forces. The present investigation was motivated by the goal to develop quantitative models of unusual behavior of hematite particle suspensions seen in experiments.

Acknowledgements M.B. acknowledges financial support from PostDocLatvia grant No. 1.1.1.2/ VIAA/3/19/562. A.C. and O.P. were supported by the Grant of Scientific Council of Latvia No.lzp-2020/1-0149.

References

1. F.J. Morin, Magnetic susceptibility of αFe_2O_3 and αFe_2O_3 with added titanium, Phys. Rev. **78**, 819 (1950) https://doi.org/10.1103/PhysRev.78.819.2
2. O. Özdemir, D.J. Dunlop, T.S. Berquó, Morin transition in hematite: size dependence and thermal hysteresis. Geochem. Geophys. Geosyst. **9** (2008) https://doi.org/10.1029/2008GC002110
3. I. Dzyaloshinsky, A thermodynamic theory of "weak" ferromagnetism of antiferromagnetics. J. Phys. Chem. Solids **4**, 241 (1958) https://doi.org/10.1016/0022-3697(58)90076-3
4. T. Moriya, Anisotropic superexchange interaction and weak ferromagnetism. Phys. Rev. **120**, 91 (1960) https://doi.org/10.1103/PhysRev.120.91
5. H.M. Lu, X.K. Meng, Morin temperature and néel temperature of hematite nanocrystals. J. Phys. Chem. C **114**, 21291 (2010). https://doi.org/10.1021/jp108703b
6. M. Brics, V. Šints, G. Kitenbergs, A. Cēbers, Energetically favorable configurations of hematite cube chains. Phys. Rev. E **105**, 024605 (2022) https://doi.org/10.1103/PhysRevE.105.024605
7. W. Lowrie, *Fundamentals of Geophysics*, 2nd edn. (Cambridge University Press, 2007). https://doi.org/10.1017/CBO9780511807107
8. L. Rossi, *Colloidal Superballs*, Ph.D. thesis,Utrecht University, Utrecht, Nederlands (2012)
9. H. Massana-Cid, F. Martinez-Pedrero, A. Cebers, P. Tierno, Orientational dynamics of fluctuating dipolar particles assembled in a mesoscopic colloidal ribbon. Phys. Rev. E **96**, 012607 (2017) https://doi.org/10.1103/PhysRevE.96.012607

10. F. Martinez-Pedrero, A. Cebers, P. Tierno, Orientational dynamics of colloidal ribbons self-assembled from microscopic magnetic ellipsoids. Soft Matter **12**, 3688 (2016). https://doi.org/10.1039/C5SM02823J
11. L. Rossi, J.G. Donaldson, J.-M. Meijer, A.V. Petukhov, D. Kleckner, S.S. Kantorovich, W.T.M. Irvine, A.P. Philipse, S. Sacanna, Self-organization in dipolar cube fluids constrained by competing anisotropies. Soft Matter **14**, 1080 (2018) https://doi.org/10.1039/C7SM02174G
12. O. Petrichenko, G. Kitenbergs, M. Brics, E. Dubois, R. Perzynski, A. Cēbers, Swarming of micron-sized hematite cubes in a rotating magnetic field—experiments. J. Mag. Mag. Mater. **500**, 166404 (2020) https://doi.org/10.1016/j.jmmm.2020.166404
13. A. Kusior, K. Michalec, P. Jelen, M. Radecka, Shaped Fe_2O_3 nanoparticles—synthesis and enhanced photocatalytic degradation towards RHB. Appl. Surface Sci. **476**, 342 (2019) https://doi.org/10.1016/j.apsusc.2018.12.113
14. P. Das, B. Mondal, K. Mukherjee, Facile synthesis of pseudo-peanut shaped hematite iron oxide nano-particles and their promising ethanol and formaldehyde sensing characteristics. RSC Adv. **4**, 31879 (2014). https://doi.org/10.1039/C4RA03098B
15. J.M. Meijer, L. Rossi, Preparation, properties, and applications of magnetic hematite microparticles. Soft Matter **17**, 2354 (2021) https://doi.org/10.1039/D0SM01977A
16. W. Chen, X. Fan, M. Sun, H. Xie, The cube-shaped hematite microrobot for biomedical application. Mechatronics **74**, 102498 (2021) https://doi.org/10.1016/j.mechatronics.2021.102498
17. M. Driscoll, B. Delmotte, M. Youssef, S. Sacanna, A. Donev, P. Chaikin, Unstable fronts and motile structures formed by microrollers. Nat. Phys. **13**, 375 (2016). https://doi.org/10.1038/nphys3970
18. V. Soni, E.S. Bililign, S. Magkiriadou, S. Sacanna, D. Bartolo, M.J. Shelley, W.T.M. Irvine, The odd free surface flows of a colloidal chiral fluid. Nat. Phys. **15**, 1188 (2019). https://doi.org/10.1038/s41567-019-0603-8
19. A. Aubret, M. Youssef, S. Sacanna, J. Palacci, Targeted assembly and synchronization of self-spinning microgears. Nat. Phys. **14**, 1114 (2018). https://doi.org/10.1038/s41567-018-0227-4
20. S.I.R. Castillo, C.E. Pompe, J. van Mourik, D.M.A. Verbart, D.M.E. Thies-Weesie, P.E. de Jongh, A.P. Philipse, Colloidal cubes for the enhanced degradation of organic dyes. J. Mater. Chem. **2**, 10193 (2014). https://doi.org/10.1039/c4ta01373e
21. J. Palacci, S. Sacanna, A.P. Steinberg, D.J. Pine, P.M. Chaikin, Living crystals of light-activated colloidal surfers. Science **339**, 936 (2013). https://doi.org/10.1126/science.1230020
22. M. Brics, V. Šints, G. Kitenbergs, A. Cēbers, *Rotating hematite cube chains* (2023), arXiv2302.13978 https://doi.org/10.48550/ARXIV.2302.13978
23. M. Belovs, M. Brics, A. Cēbers, Rotating-field-driven ensembles of magnetic particles. Phys. Rev. E **99**, 042605 (2019). https://doi.org/10.1103/PhysRevE.99.042605
24. J.-M. Meijer, D.V. Byelov, L. Rossi, A. Snigirev, I. Snigireva, A.P. Philipse, A.V. Petukhov, Self-assembly of colloidal hematite cubes: a microradian x-ray diffraction exploration of sedimentary crystals. Soft Matter. **9**, 10729 (2013). https://doi.org/10.1039/c3sm51553b
25. L. Rossi, V. Soni, D.J. Ashton, D.J. Pine, A.P. Philipse, P.M. Chaikin, M. Dijkstra, S. Sacanna, W.T.M. Irvine, Shape-sensitive crystallization in colloidal superball fluids. Proc. Nat. Acad. Sci. **112**, 5286 (2015). https://doi.org/10.1073/pnas.1415467112
26. J.-M. Meijer, Preparation and characterization of colloidal cubes, in *Colloidal Crystals of Spheres and Cubes in Real and Reciprocal Space* (Springer International Publishing, Cham, 2015), pp. 73–87. https://doi.org/10.1007/978-3-319-14809-0_5
27. C. Ruan, J. Wang, M. Gao, G. Meng Zhao, The influence of structural size on thermal stability in single crystalline hematite uniform nano/micro-cubes. Mater. Chem. Phys. **183**, 158 (2016). https://doi.org/10.1016/j.matchemphys.2016.08.014
28. S. Mallesh, D. Narsimulu, K.H. Kim, High coercivity in $\alpha - Fe_2O_3$ nanostructures synthesized by surfactant-free microwave-assisted solvothermal method. Phys. Lett. A **384**, 126038 (2020). https://doi.org/10.1016/j.physleta.2019.126038
29. G.-S. Park, D. Shindo, Y. Waseda, T. Sugimoto, Internal structure analysis of monodispersed pseudocubic hematite particles by electron microscopy. J. Colloid Interface Sci. **177**, 198 (1996). https://doi.org/10.1006/jcis.1996.0021

30. U. Schwertmann, E. Murad, Effect of ph on the formation of goethite and hematite from ferrihydrite. Clays Clay Min. **31**, 277 (1983). https://doi.org/10.1346/CCMN.1983.0310405
31. E. Matijević, P. Scheiner, Ferric hydrous oxide sols: Iii. preparation of uniform particles by hydrolysis of fe(iii)-chloride, -nitrate, and -perchlorate solutions. J. Colloid Interface Sci. **63**, 509 (1978) https://doi.org/10.1016/S0021-9797(78)80011-3
32. T. Sugimoto, M.M. Khan, A. Muramatsu, Preparation of monodisperse peanut-type $\alpha - Fe_2O_3$ particles from condensed ferric hydroxide gel. Colloids Surfaces A: Physicochem. Eng. Aspects **70**, 167 (1993a). https://doi.org/10.1016/0927-7757(93)80285-M
33. V. Malik, B. Grobety, V. Trappe, H. Dietsch, P. Schurtenberger, A closer look at the synthesis and formation mechanism of hematite nanocubes. Colloids Surfaces A: Physicochem. Eng. Aspects **445**, 21 (2014) https://doi.org/10.1016/j.colsurfa.2013.12.069
34. T. Sugimoto, M.M. Khan, A. Muramatsu, H. Itoh, Formation mechanism of monodisperse peanut-type $\alpha - Fe_2O_3$ particles from condensed ferric hydroxide gel. Colloids Surfaces A: Physicochem. Eng. Aspects **79**, 233 (1993b) https://doi.org/10.1016/0927-7757(93)80178-H
35. A. Muramatsu, S. Ichikawa, T. Sugimoto, Controlled formation of ultrafine nickel particles on well-defined hematite particles. Colloids and Surfaces A: Physicochemical and Engineering Aspects **82**, 29 (1994) https://doi.org/10.1016/0927-7757(93)02592-3
36. M. Satheesh, A.R. Paloly, K. Suresh, M. Junaid Bushiri, Influence of solvothermal growth condition on morphological formation of hematite spheroid and pseudocubic micro structures and its magnetic coercivity. J. Phys. Chem. Solids **98**, 247 (2016). https://doi.org/10.1016/j.jpcs.2016.07.020
37. J.W.J. de Folter, E.M. Hutter, S.I.R. Castillo, K.E. Klop, A.P. Philipse, W.K. Kegel, Particle shape anisotropy in pickering emulsions: cubes and peanuts. Langmuir **30**, 955 (2014). pMID: 24020650, https://doi.org/10.1021/la402427q
38. H. Massana-Cid, F. Martinez-Pedrero, E. Navarro-Argemí, I. Pagonabarraga, P. Tierno, Propulsion and hydrodynamic particle transport of magnetically twisted colloidal ribbons. New J. Phys. **19**, 103031 (2017b). https://doi.org/10.1088/1367-2630/aa84f9
39. See supplemental material at [https://doi.org/10.1007/978-3-031-58376-6_6] for further videos

Chapter 7
The Synchronous to Exchange Transition in Magnetically Driven Colloidal Dimers

Mattia Ostinato, Antonio Ortiz-Ambriz, and Pietro Tierno

7.1 Introduction

Microscopic colloidal particles under time-dependent external fields represent an accessible model system to investigate the fascinating emergent dynamics that occur when many-body systems are driven out of equilibrium [1–5]. Colloids have a size in the visible wavelength, are characterized by experimentally accessible time scales, and can be easily manipulated with the aid of relatively low intensity external fields [6–8]. When the particles are located close to a wall, or within a narrow channel, the combination between pair interactions and confinement may lead to novel dynamics and emerging phenomena [9–13].

Here, we use numerical simulations to investigate the dynamic states emerging from a collection of magnetic particles strongly confined between two thin plates such that overpassing along the perpendicular direction is forbidden. A similar confinement was studied in the past to model geometric frustration, but it used size-tunable hydrogel particles without an external field, and thus the particles were passive and

M. Ostinato · A. Ortiz-Ambriz · P. Tierno (✉)
Departament de Física de la Matèria Condensada, Universitat de Barcelona, 08028 Barcelona, Spain
e-mail: ptierno@ub.edu

M. Ostinato
e-mail: mattia.ostinato@ub.edu

A. Ortiz-Ambriz
e-mail: aortiza@tec.mx

Universitat de Barcelona Institute of Complex Systems (UBICS), Universitat de Barcelona, Barcelona, Spain

A. Ortiz-Ambriz · P. Tierno
Institut de Nanociència i Nanotecnologia, Universitat de Barcelona, Barcelona, Spain

A. Ortiz-Ambriz
Escuela de Ingeniería y Ciencias, Tecnológico de Monterrey, 64849 Monterrey, Mexico

© The Author(s) 2024 69
B. Doudin et al. (eds.), *Magnetic Microhydrodynamics*, Topics in Applied Physics 120,
https://doi.org/10.1007/978-3-031-58376-6_7

non driven [14–16]. Instead we consider the situation where, in the presence of a time-dependent field, the particles form a series of dynamic states resulting from the combination of excluded volume, confinement and induced magnetic dipolar forces. These states were recently realized experimentally [17, 18], and here we present numerical simulation results aiming at investigating the transition between two of them. In particular, we focus on the transition between a collection of synchronously rotating, localized dimers and an exchange phase, where the dimers break and exchange particles between them. We start the present contribution by illustrating the experimental system and the different dynamic states that the particles form by varying the field parameters. After that, we describe in detail the numerical simulation scheme adopted. Later we define our order parameter and how we extract different information on the nature of the observed transition. We finally conclude the manuscript by resuming the main results and discussing the nature of the observed transition.

7.2   Realization of the Colloidal Dimers and Exchange States

7.2.1   The Experimental System

The experimental system was developed in Ref. [17] and employed commercial paramagnetic colloids (Dynabeads M-270) made of a cross-linked polystyrene matrix with surface carboxylic groups. These particles have an average diameter $d = 2.8\,\mu$m and present a narrow size distribution. The magnetic properties of these particles arise from the uniform doping with iron oxide superparamagnetic grains ($\sim 20\%$ by vol.) which increases the particle density to $\rho = 1.6\,\mathrm{g\,cm^{-3}}$. Due to this doping, the particles can be controlled by an external magnetic field $\boldsymbol{B}$. In particular, when $\boldsymbol{B} \neq 0$, the paramagnetic colloids acquire an induced dipole moment which points along the field direction, $\boldsymbol{m} = \pi d^3 \chi \boldsymbol{B}/(6\mu_0)$, being χ the magnetic volume susceptibility of the particle, and μ_0 the permeability of vacuum. Thus, a pair of particles (i, j) at a distance $r = |\boldsymbol{r}_i - \boldsymbol{r}_j|$ interacts via the magnetic dipolar potential,

$$U_{dip} = -\frac{\mu_0}{4\pi r^5} \left[3(\boldsymbol{m}_i \cdot \boldsymbol{r})(\boldsymbol{m}_j \cdot \boldsymbol{r}) - (\boldsymbol{m}_i \cdot \boldsymbol{m}_j)r^2 \right] \tag{7.1}$$

which is attractive (repulsive) for particles with magnetic moments parallel (perpendicular) to r.

The particles were diluted in highly deionized water, and confined between two glass surfaces made of a plain microscope slide and a coverslip. To achieve a small confinement, both plates were manually pressed, and later glued with a fast-curing epoxy adhesive. With this method, it was possible to obtain a small thickness in the range $h \in [3, 6]\mu$m. Such thickness was measured by analyzing the horizontal projection length of formed dimers under a static, perpendicular field $\boldsymbol{B} = B\boldsymbol{z}$. The particles were visualized with an upright optical microscope which was equipped with a set of custom build magnetic coils, that allows generating homogeneous, static and time-dependent magnetic fields.

7.2.2   Colloidal States Under the Precessing Field

From Eq. 7.1 follows that the pair interactions between the paramagnetic colloids can be tuned by an external field B. When the particles are confined above a plane, a static field $B = Bz$ applied perpendicular to such plane induces an isotropic repulsion and, within the correct range of B, one can induce the formation of a triangular lattice [19, 20]. In contrast, anisotropic attractive interactions can be induced via a static, in-plane field applied along the x or y axes [21, 22]. This situation becomes different when the particles are confined between two plates, as shown in Fig. 7.1d. Osterman et al. [23] demonstrated that the confinement may soften the pair repulsion. For particles enclosed between two hard walls and separated by a distance $h < 2d$, the potential in Eq. 7.1 may be rewritten as,

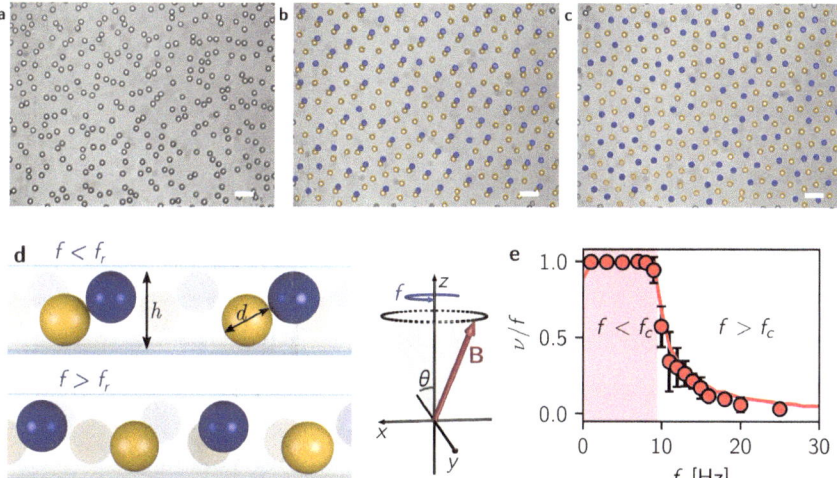

Fig. 7.1   a–c Optical microscope images of confined paramagnetic colloids within a cell of thickness $h = 3.9\,\mu$m. In **a** there is no external field, in **b** the particles are subjected to a precessing field with amplitude $B_0 = 7.3$ mT, cone angle $\theta = 26.9°$, frequency $f = 1$ Hz, and in **c** the frequency is raised to $f = 20$ Hz. All images have a scale bar of $10\,\mu$m. **d** Left: Sketch of the particles inside a cell of thickness $d < h < 2d$ for frequency $f < f_r$ (top) and $f > f_r$ (bottom). Right: The precessing magnetic field with the cone angle θ. **e** Normalized angular frequency v of a dimer versus field frequency f showing the transition from synchronous to asynchronous regime at $f_c = 9.8$Hz. Scattered data are experimental results, the continuous line is from numerical simulation. Image adapted with permission from Ref. [17].

$$U_{dip} = -\frac{\mu_0 m^2}{4\pi} \left[\frac{r^2 - 2z^2}{(r^2 - z^2)^{5/2}} \right], \quad (7.2)$$

which shows that two particles repel when the elevation difference between their centers Δz, is $\Delta z < d/\sqrt{5}$, and otherwise they experience a short-range attractive and long-range repulsive potential. Such interactions give rise to different self assembled structures at equilibrium, featuring hexagonal, square, stripe or labyrinth-like ordering.

In contrast to a static field, the formation of interacting dimers was induced via a time-dependent field, as shown in Fig. 7.1d. This field performs a conical rotation around an axis z perpendicular to the sample plane with a frequency f and a cone angle θ,

$$\mathbf{B} = B_0[\cos\theta\hat{\mathbf{z}} + \sin\theta(\cos(2\pi ft)\hat{\mathbf{x}} + \sin(2\pi ft)\hat{\mathbf{y}})], \quad (7.3)$$

where B_0 is the field amplitude. Under this type of forcing, novel dynamic colloidal patterns were observed. In particular, the sequence of images in Fig. 7.1a, c shows how a colloidal suspension confined to a narrow cell of thickness $h = 3.9\,\mu m$ self-organizes from an initial disordered phase (a) with $\mathbf{B} = 0$. A precessing field with a relative slow frequency ($f = 1\text{Hz}, \theta = 26.9°$) arranges the paramagnetic colloids into an ensemble of rotating dimers, which perform a rotation around the z axis (b). Each dimer is composed of two particles, one closer to the top plate ("up") and the other closer to the bottom one ("down"). They can be experimentally distinguished by their different brightness resulting from the different elevations, and are highlighted in the images by two colors. The dimers are stable as long as the field is kept fixed, and this state can be destabilized by increasing f. As shown in Fig. 7.1c, by raising the driving frequency to $f = 20\text{Hz}$, the dimers break and the colloidal systems transforms into two separated lattices made of up and down particles. The up particles are close to the top plane, and remain there as long as the field is applied. A corresponding lateral sketch of the the particle locations with respect to the plates can be found in Fig. 7.1d. Depending on the density of the particles and the cell thickness, a continuous variation of f can drive the system into two different high frequency states. For each transition path, a transition frequency can be defined, namely f_r, which separates the stable dimer from the broken (up and down) state and a synchronous to asynchronous transition frequency, f_c. The latter separates two different rotational modes of the dimers, Fig. 7.1e. When $f < f_c$, the dimers rotate synchronously with the precessing field, and their rotational frequency is $\nu = f$. In contrast, for $f > f_c$, the phase-lag angle between $\mathbf{B}$ and the dimer long axis changes in time, and the dimers enter into an asynchronous regime, showing a characteristic "back-and-forth" rotation. Due to these oscillations, their rotational motion decreases, as shown in Fig. 7.1e.

The different dynamic states depend on both the frequency f and cell thickness h. As shown in the schematic in Fig. 7.2a, we find four types of transition paths when starting from the synchronous regime, i.e. at low frequency. The first is the Synchronous-Asynchronous (SA), which can be more complex with an intermediate Exchange state (SEA). In such state which is usually observed at high density, the

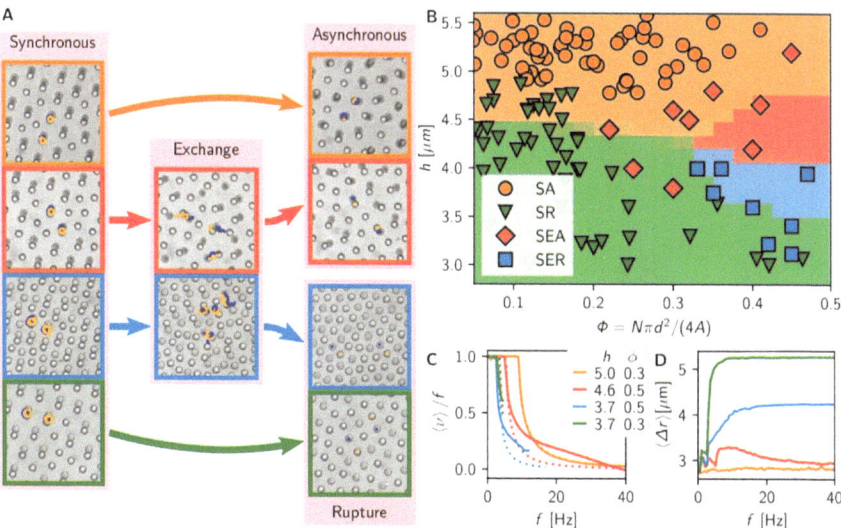

Fig. 7.2   a Four transition paths, experimentally observed under a field of amplitude $B_0 = 7.28$ mT and cone angle $\theta = 26.9°$. In all cases the starting frequency is $f = 1$ Hz. The first image on the top illustrates the Synchronous $\rightarrow$ Asynchronous (SA) transition ($f = 20$ Hz) with a cell thickness of $h = 5.1\,\mu$m. The second the Synchronous $\rightarrow$ Exchange ($f = 8$ Hz) $\rightarrow$ Asynchronous (SEA) transition ($f = 25$ Hz) $h = 4.4\,\mu$m, the third, the Synchronous $\rightarrow$ Exchange ($f = 3$ Hz) $\rightarrow$ Rupture (SER) of dimers ($f = 14$ Hz), $h = 4.4\,\mu$m. The last, the Synchronous $\rightarrow$ Rupture (SR) transition ($f = 9$ Hz), $h = 4\,\mu$m. **b** Regions where the transition paths occurs in the (Φ, h) plane. Scattered symbols are experimental data, shaded regions result from numerical simulations. **c** Mean rotation frequency versus f and **c** nearest neighbor separation distance $\langle\Delta r\rangle$ for different cell thickness h and normalized area packing fraction Φ. Image adapted with permission from Ref. [17]

dimers break up, and the composing particles exchange positions by passing close to near particles, forming a rotating dimer for half period of the field, and then breaking again. Further, the synchronous-exchange path may instead be followed by a rupture state (SER), namely the up and down crystal or simply the synchronous state can go directly into the rupture one (SR). The diagram in Fig. 7.2b illustrates their locations when varying the normalized area packing fraction, $\Phi = N\pi d^2/(4A)$, being N the number of particles and A the corresponding area. One can identify these states by measuring two observables related with the particle dynamics. They are the average rotation speed of the dimers, shown in Fig. 7.2c, and $\langle\Delta r\rangle$ which is the average distance between nearest neighbors, Fig. 7.2d. In the numerical work, we focus on the transition between the synchronous-exchange state along the SER path, and carefully analyze how this transition set in by raising f.

7.3 Numerical Simulation

We perform Brownian dynamics simulations using the free package LAMMPS [24] modified to consider the particle induced dipole moment and an overdamped integrator. In particular, we simulate $N = 1000$ paramagnetic colloids with individual positions $r_i = (x_i, y_i, z_i)$ confined in a quasi two dimensional (2D) box of size $L_x \times L_y \times h$. The box has periodic boundary conditions on the (x, y) plane and fixed walls on the z axis at positions $z = \pm \frac{h}{2}$. For each particle, we integrate the equations of motion,

$$\gamma \frac{dr_i}{dt} = \sum_{j \neq i} F_{int}(r_i - r_j) + F_w + F_g + \eta(t) \tag{7.4}$$

where γ is the viscous friction, $F_{int}(r_i - r_j)$ is the total force exerted on particle i by particle j, F_w is the normal force exerted by the confining walls on particle i, F_g the gravitational force and $\eta(t)$ the force due to the thermal fluctuations. The total force on the particle results from an interaction potential, $F_{int}(r_i - r_j) = -\nabla U_{int}(r_i - r_j)$, where $U_{int}(r_i - r_j) = U_{dip}(r_i - r_j) + U_{WCA}(|r_i - r_j|)$. Here the first term is the magnetic dipolar interaction (Eq. 7.1), while the second one refers to a repulsive Weeks-Chandler-Andersen (WCA) potential U_{WCA}, which is given by,

$$U_{WCA} = \begin{cases} 4\epsilon \left[\left(\frac{d}{r} \right)^{12} - \left(\frac{d}{r} \right)^{6} \right] + \epsilon & \text{for } r < 2^{\frac{1}{6}}d \\ 0 & \text{for } r > 2^{\frac{1}{6}}d \end{cases} \tag{7.5}$$

Further, the interaction between the particles and the wall, $U_w(z)$ is also given by a WCA potential, with $F_w(z) = -\nabla U_w(z)$ and the gravitational force is given by $F_g = -\pi \Delta \rho g d^3 z / 6$ being $\Delta \rho$ the density mismatch between the particles and water and g the gravitational acceleration. Finally we assume that $\eta(t) \equiv (\eta_x, \eta_y, \eta_z)$ are random Gaussian variables with zero mean, $\langle \eta_i(t) \rangle = 0$ and correlation function: $\langle \eta_i(t)\eta_j(t') \rangle \equiv 2k_B T \gamma \delta_{ij} \delta(t - t')$, being k_B the Boltzmann constant and T the experimental temperature.

In the simulations, we usually fix N and the normalized area fraction $\Phi = \frac{N\pi d^2}{4A}$ being $A = L_x \times L_y = L^2$ and varying the driving frequency f, which is our control parameter. For each value of the frequency, $M = 10$ statistically independent runs are performed, each with a randomized initial configuration. The obtained observable are then averaged over these M configurations. We introduce parameters extracted directly from the experiments, such as $\phi = 0.262$, $h = 3.9$ μm, $B_0 = 7.28$ mT, $d = 2.8 \mu$m, $\theta = 27°$, $\gamma = 56.75 \times 10^{-6}$ pN s nm^{-1}, $\Delta \rho = 10^3$ Kg m^{-3}, $\chi = 0.4$ and use as value for the WCA potential $\epsilon = 10^4$ pN nm [25, 26]. Our system is initialized by randomly placing the particles in the $z = 0$ plane, with the values of the coordinate of each particle center, x_i and y_i being drawn each independently from a uniform distribution in $(-\frac{L}{2}, \frac{L}{2})$. We further avoid artifacts due to the close proximity of the particles by imposing that each particle position must be at a minimum distance of $d + \delta$, being δ a tolerance parameter which is set to 50 nm. Equations 7.4

are then integrated with a simulation time step of $\delta t = 10^{-4} s$ and a total simulation time of $t_{tot} = 1000\,s$.

Finally, in the runs that investigate the SER transition path, we switch off gravity to avoid perturbing the final distribution of the up and down particles.

7.4 The Synchronous-Exchange Transition

7.4.1 Order Parameter

We use numerical simulations to investigate the SE transition which occurs when the dimers break and exchange particles with their neighbors. The advantage of the simulation is that it allows us to carefully tune the driving frequency and to consider relatively large systems, increasing the statistical average. As an order parameter that allows to distinguish between the synchronous and the exchange state, we use a combination of stroboscopic measurements and Voronoi tessellation, following a previous work [27]. In particular, our procedure is schematically illustrated in Fig. 7.3a–d. First, for a state at time t we define the Voronoi tessellation as the set of polygons $\{a_i(t)\}$ for $i \in \{1, 2, ..., N\}$ such that the area inside the polygon $a_i(t)$ contains only the points whose (Euclidean) distance is closer to particle i than other particles. Then, a particle is defined as *active* at time t if its position after half a period, $r_i(t + T/2)$, is contained by its Voronoi polygon half a period before, $a_i(t - T/2)$. The series of images in Fig. 7.3a–d show how this definition applies to the synchronous (a, b) and exchange (c, d) states. In the former regime, the particles within the dimers at $t = t' - T/2$ (a) display similar positions after a period, $t = t' + T/2$ (b), and identical Voronoi tessellation (green mesh). This fact results from the reversible trajectories which cyclically repeat after one period even for isolated particles which in this regime do not perform hopping motion, with negligible deviation due to thermal fluctuations. Thus, all particles in Fig. 7.3, panels (a, b) are considered as passive. The situation changes in the exchange phase, illustrated in panels (c, d). After one driving period, at $t = t' + T/2$ (d), due to the exchange position process, many particles are no longer in their original cells, and this generates *active* particles, which are highlighted by an orange ring.

Once identified the active particles, we define the order parameter $\Psi(t, f)$ as the fraction of such particles at time t; $\Psi(t, f) = N_{active}(t, f)/N$, being N the total number of particles. In Fig. 7.3e we show the time evolution of Ψ averaged over 10 simulations for $N = 1000$ particles, starting from random initial conditions and in the range of frequencies $f \in [3.1, 3.3]\,\text{Hz}$. Here it should be noted that single isolated dimers display a transition from synchronous rotation to broken dimer above 3.3 Hz, sign of the collective nature of the exchange phase. Moreover, since the range of frequencies is very small, all curves in Fig. 7.3e start from similar initial values. However, after an initial relaxation time, different frequencies produce very differ-

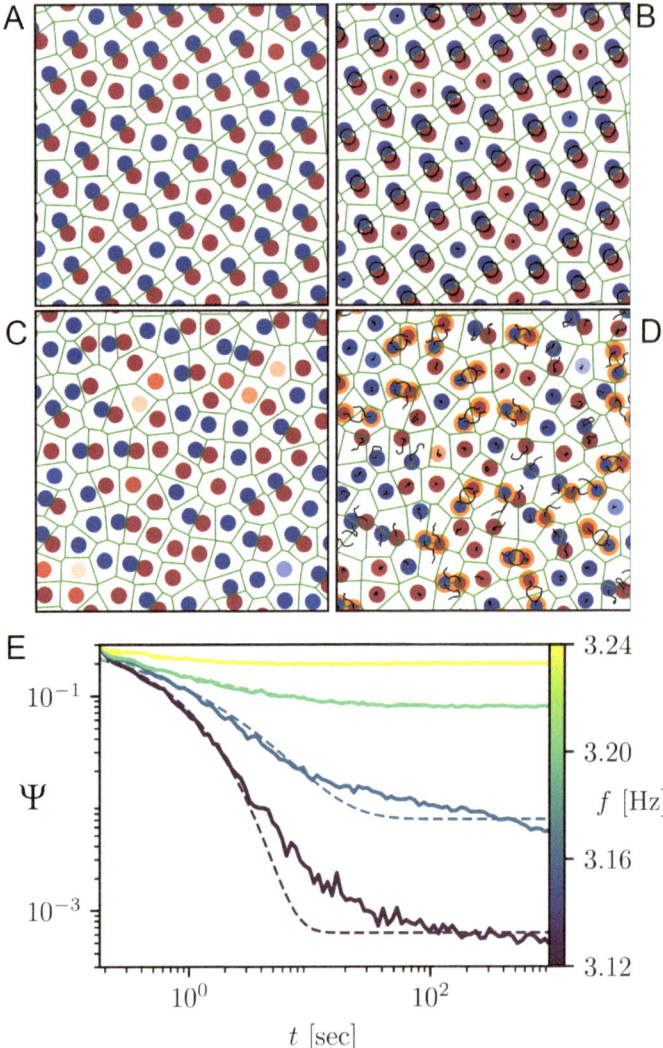

Fig. 7.3 **a–d** Simulation snapshots explaining the classification of particles into active and passive. Top **a, b** panels corresponds to $f = 2.8$ Hz while bottom **c, d** to $f = 4$ Hz. Left **a, c** panels are taken at time $t = t' - T/2$, while right panels **b, d** at $t = t' + T/2$, being T the period of rotation of the magnetic field. The particles colored in red (blue) are close to (far away from) the top wall. The green mesh illustrates the Voronoi tessellation, while the particle trajectories are superimposed to the images in black in panels **b, d**. The orange disks in panel **d** indicates colloidal particles that are considered "active". **e** Evolution with time of the fraction of active particles Ψ for four different driving frequencies averaged from 10 simulation runs ($N = 1000$). The continuous lines through the data are non linear regressions following a stretched exponential, see text

ent final steady-state values of activity. During the relaxation process, we observe that colloidal particles close enough were able to join forming dimers. At high frequencies, these dimers are not stable but keep breaking and reforming, impeding the system to leave the active state. However, at low frequencies, the dimer can survive long enough to interact repulsively with neighboring dimers and arrange themselves into a low ordered configuration, like that depicted in Fig. 7.3a. In such state, the dimers are mainly stable, and unpaired colloids are isolated, with no possibility of hopping, thus reaching a low activity state.

We find that all curves in Fig. 7.3e can be fitted by a stretched exponential function of the form,

$$\Psi(t) = \Psi_0 \exp\left[-\left(\frac{t}{\lambda}\right)^{\beta}\right] + \Psi_\infty \tag{7.6}$$

with a stretching exponent $\beta \in [0, 1]$. This functional form is commonly used when modeling relaxation in glasses, or to approximate response functions (be it mechanical, electric or magnetic) in disordered media.

7.4.2 Relaxation Time

Using Eq. 7.6, we could fit all the simulation data, and extract two main parameters, the steady-state activity from the constant $\Psi|_{t\to\infty} = \Psi_0$ and the relaxation time scale which is given by the first moment of the stretched exponential function [28]:

$$\tau_r = \frac{\lambda}{\beta} \Gamma\left(\frac{1}{\beta}\right), \tag{7.7}$$

being Γ the gamma function. Figure 7.4a, b show both quantities as a function of the reduced frequency, here defined as $|f - f_r|/f_r$ being f_r the rupture frequency that bridges the synchronous to the exchange phase. To determine f_r, we start by considering the behavior of the asymptotic activity, which is shown in the inset of Fig. 7.4a. While the left of the plot is nearly flat, with a small increase which could be due to thermal fluctuations, $\Psi|_{t\to\infty}$ rapidly raises for $f > f_r$ following a power law behavior. We fit this plot using a piece-wise function of the form $((f - f_r)/f_r)^{\gamma}$ for $f > f_r$ and 0 otherwise, and extract a value for the critical frequency of $f_r = 3.185 \pm 0.001$ Hz, with an exponent $\gamma = 0.7$. Using the value of f_r, we can calculate the reduced frequency as $|f - f_r|/f_r$ and we use it to plot the one branch of the asymptotic activity in Fig. 7.4a. In Fig. 7.4b, we show the relaxation timescale τ_r extracted from the two branches close to f_r as a function of the reduced frequency, while the small inset shows the full region around f_r in linear scale. Here the right branch $(f > f_r)$ is displayed as a solid line, while the left branch $(f < f_r)$ as a dotted one. While these data are noisier than the asymptotic activity, they still exhibit a linear behavior in the logarithm plot. We find that both curves are consistent with the 1.23

Fig. 7.4 **a** Asymptotic value of the order parameter Ψ in the long time limit against reduced frequency. Here f_r denotes the critical frequency of rupture, calculated from fitting the plot in the inset to a piece-wise power law, which gives an exponent $\gamma = 0.7$ (orange lines). **b** Relaxation time of Ψ after a random initial state. Both branches appear to be consistent with an exponent of 1.23 (dotted line). The inset shows the divergence of the timescale around the critical frequency f_r

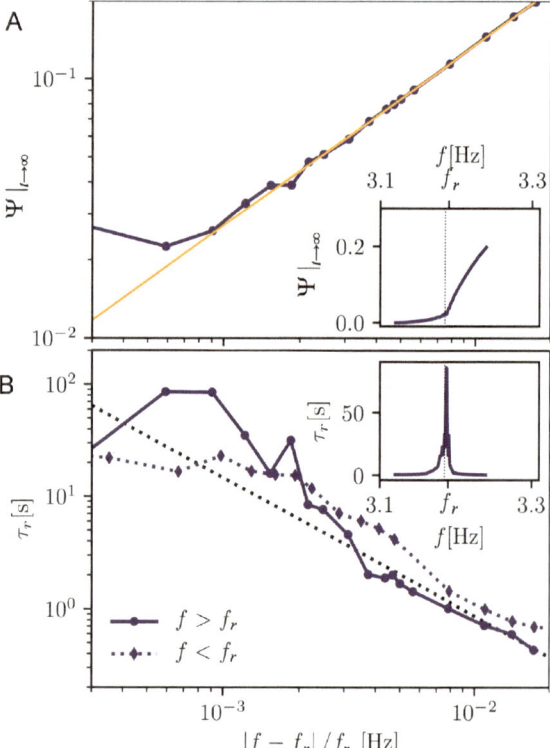

exponent predicted for the Manna (conserved-DP) universality class [29], which is shown as a dotted line.

7.5 Conclusion

We have investigated the collective dynamics of paramagnetic colloids driven by a precessing field while being confined within two narrow plates. We focus on the transition between two dynamic states, synchronous and exchange, which occurs at large particle densities. We identify a critical rupture frequency that bridges these two states and define an order parameter as the fraction of active particles. We find that the time dependence of the order parameter can be well capture by a stretched exponential function which allows us to measure the relaxation time associated with such transition. We find an exponent consistent with the Manna universality class, which points toward the presence of an absorbing phase transition in such system.

Out of equilibrium phase transitions have been reported in models that describe disparate phenomena, from the onset of turbulence [30], to forest fires [31] and financial crises [32]. Among these, a recurrent observation is the existence of an absorbing phase, which appears in systems that fall into a state from which they can't get out [33]. Other examples of absorbing phases are observed in models for catalytic chemical reactions [34] interface growth [35], wetting [36], depinning [37], and granular matter [38]. Also recently, periodically sheared emulsions have risen as model systems for non-equilibrium transitions with an absorbing phase; one of the very few experimental realizations of such a transition [39, 40]. We have provided another system that can be used to investigate the fascinating collective physics of these systems when they are driven out of equilibrium by an external field.

Acknowledgements We thank Helena Massana-Cid for experimental advice and Hugues Chaté for stimulating discussions. The authors acknowledge support from the European Research Council (grant agreement no. 811234). P.T. acknowledges support from the Generalitat de Catalunya under the program "ICREA Acadèmia."

References

1. H. Löwen J. Phys. Condens. Matter **20**, 404201 (2008)
2. E. Martin, A. Snezhko, Rep. Prog. Phys. **76**, 126601 (2013)
3. J. Dobnikar, A. Snezhko, A. Yethiraj, Soft Matter **9**, 3693–3704 (2013)
4. S.H.L. Klapp, Curr. Opin. Colloid Interface Sci. **21**, 76 (2016)
5. P. Tierno, A. Snezhko. ChemNanoMat **7**, 881 (2021)
6. H. Löwen, Phys. Rep. **237**, 249–324 (1994)
7. W. Poon, Science **304**, 830–831 (2004)
8. D. Babič, C. Schmitt, C. Bechinger, Chaos **15**, 026114 (2005)
9. A. Snezhko, I.S. Aranson, Nat. Mater. **10**, 698–703 (2011)
10. A. Cebers, Curr. Op. Coll. Int. Sci. **10**, 167–175 (2011)
11. P. Tierno, T.M. Fischer, Phys. Rev. Lett. **112**, 048302 (2014)
12. R. Abedini-Nassab, D.Y. Joh, M. Van Heest, C. Baker, A. Chilkoti, D.M. Murdoch, B.B. Yellen, Adv. Funct. Mater. **26**, 4026–4034 (2014)
13. M. Driscoll, B. Delmotte, Curr. Opin. Coll. Int. Sci. **40**, 42–57 (2019)
14. Y. Han, Y. Shokef, A.M. Alsayed, P. Yunker, T.C. Lubensky, A.G. Yodh, Nature **456**, 898–903 (2008)
15. Y. Shokef, T.C. Lubensky, Phys. Rev. Lett. **102**, 048303 (2009)
16. D. Zhou, F. Wang, B. Li, X. Lou, Y. Han, Phys. Rev. X **7**, 021030 (2017)
17. H. Massana-Cid, A. Ortiz-Ambriz, A. Vilfan, P. Tierno, Sci. Adv. **6**, eaaz225 (2020)
18. F. Meng, A. Ortiz-Ambriz, H. Massana-Cid, A. Vilfan, R. Golestanian, P. Tierno, Phys. Rev. Res. **2**, 012025 (2020)
19. K. Zahn, R. Lenke, G. Maret, Phys. Rev. Lett. **82**, 2721 (1999)
20. W. Wen, L. Zhang, P. Sheng, Phys. Rev. Lett. **85**, 5464–5467 (2000)
21. A.T. Skjeltorp, Phys. Rev. Lett. **51**, 2306–2309 (1983)
22. R. Toussaint, G. Helgesen, E.G. Flekkøy, Phys. Rev. Lett. **93**, 108304 (2004)
23. N. Osterman, D. Babič, I. Poberaj, J. Dobnikar, P. Ziherl, Phys. Rev. Lett. **99**, 248301 (2007)
24. S. Plimton, J. Comput. Phys. **117**, 1–19 (1995)
25. A. Ortiz-Ambriz, P. Tierno, Nat. Commun. **7**, 10575 (2016)
26. C. Rodríguez-Gallo, A. Ortiz-Ambriz, P. Tierno, Phys. Rev. Lett. **126**, 188001 (2021)
27. J.H. Weijs, R. Jeanneret, R. Dreyfus, D. Bartolo, Phys. Rev. Lett. **115**, 108301 (2015)
28. D.C. Johnston, Phys. Rev. B **74**, 184430 (2006)
29. S.S. Manna, J. Phys. A: Math. Gen. **24**, L363 (1991)

30. P. Rupp, R. Richter, I. Rehberg, Phys. Rev. E **67**, 036209 (2003)
31. E.V. Albano, J. Phys. A: Math. Gen. **27**, L881 (1994)
32. W. Weidlich, Phys. Rep. **204**, 1 (1991)
33. H. Hinrichsen, Adv. Phys. **49**, 815 (2000)
34. R.M. Ziff, E. Gulari, Y. Barshad, Phys. Rev. Lett. **56**, 2553–2556 (1986)
35. M. Kardar, G. Parisi, Y.-C. Zhang, Phys. Rev. Lett. **56**, 889–892 (1986)
36. H. Hinrichsen, R. Livi, D. Mukamel, A. Politi, Phys. Rev. Lett. **79**, 2710 (1997)
37. M. Kardar, Phys. Rep. **301**, 85 (1998)
38. C. Ness, M.E. Cates, Phys. Rev. Lett. **124**, 088004 (2020)
39. D.J. Pine, J.P. Gollub, J.F. Brady, A.M. Leshansky, Nature **438**, 997 (2005)
40. L. Corté, P.M. Chaikin, J.P. Gollub, D.J. Pine, Nat. Phys. **4**, 420 (2008)

Part III
Water and Solutions

Chapter 8
Influence of Magnetic Field on Water and Aqueous Solutions

Sruthy Poulose, Jennifer A. Quirke, and Michael Coey

8.1 Introduction

The influence of a magnetic field B (measured in Tesla) on condensed matter depends on the nature of the field, whether homogeneous or inhomogeneous, static or dynamic. The simultaneous application of an electric field E (measured in Vm^{-1}) creates a dielectric polarization or excites an electric current that can modify the magnetic response. In this chapter, our focus is on the response of water and aqueous solutions to homogeneous and inhomogeneous static magnetic fields. B is the fundamental divergenceless magnetic field with no sources or sinks. The magnetic response of condensed matter however is determined by a different field, the local magnetic field strength H. The two are related by the equation,

$$B = \mu_0(H + M) \tag{8.1}$$

where μ_0 is the magnetic constant $4\pi\ 10^{-7}\ TmA^{-1}$, and the magnetization M is the magnetic moment *per unit volume* of condensed matter. Units of H and M are both Am^{-1}; an equivalent unit for M is $JT^{-1}\ m^{-3}$, based on the expression for the energy of a magnetic moment m (units Am^2) in a field of B, $E = -m.B\ JT^{-1}$.

S. Poulose · J. A. Quirke · M. Coey (✉)
School of Physics, Trinity College, Dublin, Ireland
e-mail: jcoey@tcd.ie

J. A. Quirke
e-mail: jequirke@tcd.ie

© The Author(s) 2024
B. Doudin et al. (eds.), *Magnetic Microhydrodynamics*, Topics in Applied Physics 120,
https://doi.org/10.1007/978-3-031-58376-6_8

8.1.1 Susceptibility

The basic response of water or aqueous solution to a field H is the appearance of an induced magnetization, proportional to the magnetic susceptibility χ of the liquid, defined by

$$M = \chi H \tag{8.2}$$

M is isotropic and initially linear in field. Thus defined, susceptibility is a pure number, with no dimensions. The value for water is -9.0×10^{-6}, so water is diamagnetic with induced magnetization directed opposite to the applied field H_0. The local magnetic field strength is related to H_0 via a 'demagnetizing' factor that depends only on the shape of the sample with limits $0 \leq \mathcal{N} \leq 1$.

$$H = H_0 - \mathcal{N}M \tag{8.3}$$

Dividing Eq. 8.3 by H, it can be seen that H is actually bigger that H_0 when χ is negative, but the difference for water is very small, and it can usually be neglected.

Other definitions of susceptibility are possible, where it is not dimensionless. Sometimes H_0 is replaced by B_0. Also cgs units (emu, Oersted and Gauss) may be encountered. The dimensionless cgs susceptibility is smaller than its SI counterpart by a factor 4π because $\mu_0 = 1\text{GOe}^{-1}$ in the cgs system. Table 8.1 lists values for the susceptibility of water, together with its units, for different definitions to ease confusion when reading literature that does not use Eq. 8.1 and SI units. Besides the magnetic moment per unit volume used in our definition of χ, the magnetic moment per unit mass or per mole may be used instead. There is little temperature dependence for water; the susceptibility is found to increase by 1% in the range 1–80 °C due to a slight weakening of the hydrogen bonds [12, 48].

There are two notable magnetic consequences of the diamagnetic susceptibility of pure water in a non-uniform field. One is the *Moses Effect* [5], where a field is applied to region of water of depth d in an open bath. When a long bath of water is

Table 8.1 Summary of different ways of defining the susceptibility of water, with corresponding units in brackets

Field	Volume susceptibility	Mass susceptibility[a,b]	Molar susceptibility[c]
H (Am^{-1})	-9.0×10^{-6} (–)	-9.0×10^{-9} (m^{3}kg^{-1})	-1.62×10^{-10} (m^{3}mol^{-1})
B (T)	-7.2 (JT^{-2} m^{-3})	-7.2×10^{-3} (JT^{-2} kg^{-1})	-1.29×10^{-4} (JT^{-2} mol^{-1})
H (Oe) B (G)	-7.2×10^{-7} (–)	-7.2×10^{-7} (emu Oe^{-1} g^{-1})	-1.29×10^{-5} (emuOe^{-1} mol^{-1})

A dash indicates a susceptibility that is dimensionless
[a] 1 emu $= 10^{-3}$ Am^2. [b] 1G $= 10^{-4}$ T. [c] 1 mol of water $= 0.018$ kg

placed between the poles of an electromagnet, a small depression δ of the surface in the field is observed due to the magnetic pressure

$$P_{\mathrm{m}} = \frac{1}{2}\mu_0\chi H^2 \tag{8.4}$$

exerted on the water. Equating this to the change in hydrostatic pressure $\delta\rho g$. The value of δ for $\mu_0 H = 1$ T is just 360 μm. In 10 T, the effect rises to 3.6 cm. (The red sea is 3 km deep)

The other effect is *diamagnetic levitation*. A pendant droplet in a vertical magnetic field gradient $\nabla_z B$ experiences an apparent change of weight due to the Kelvin force density, also known as the 'magnetic field gradient force' density $F_K = -(1/2\mu_0)\chi \nabla_z B^2$. The force will counterbalance the weight when $-(1/\mu_0)B\nabla_z B = -\rho g$, where ρ is the density of water. The numerical condition for levitation of water is therefore

$$B\nabla_z B = -1360\,\mathrm{T}^2\mathrm{m}^{-1}. \tag{8.5}$$

It is possible to suspend water droplets and other objects composed mainly of water (frogs, strawberries ....) in stable equilibrium close to the top of an open bore of a Bitter magnet or superconducting solenoid where the field is nonuniform and ~ 20 T [56].

A further effect on a pendant or suspended water droplet is due to *Maxwell stress*. An applied field $\boldsymbol{H}_0$ induces a small uniform magnetization $\boldsymbol{M}$ in droplets of isotropic paramagnetic or diamagnetic liquids, as indicated in Fig. 8.1. The 'demagnetizing' effect can usually be neglected in water on account of its small susceptibility, although this is not the case in strongly paramagnetic ferrofluids with $\chi \sim 1$, which minimize their total energy in the magnetic field by adopting spectacular spiked surfaces [54]. The pressure is not continuous at an interface between a liquid magnetized in a uniform field and vacuum, where a 'magnetic normal traction'

$$P_{\mathrm{n}} = \frac{1}{2}\mu_0 M_\perp^2 \tag{8.6}$$

acts perpendicular to the interface [16, 54].

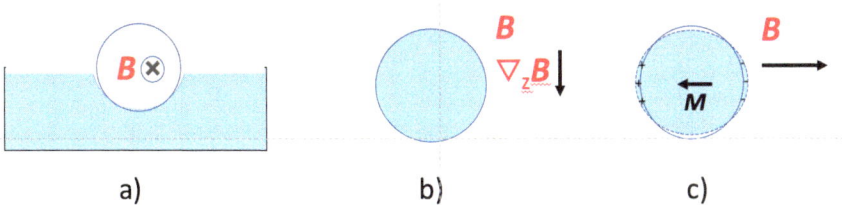

Fig. 8.1 Effects of a magnetic field on water. **a** Moses effect in a localized field **b** levitation in a large vertical field gradient and **c** Maxwell stress on a spherical droplet in a uniform field

8.2 Pendant Droplets

Fitting the shape of pendant liquid droplets captured in an optical contact analyser is a commonly used method to determine the surface tension of liquids. The pear-shape of the droplet is a consequence of equilibrium between gravitational and surface tension forces. Commercial optical droplet analysers fit the shape of the outline of a digital photographic image of the pendant droplet to a parametric equation from which surface tension γ is determined, given the liquid density ρ. Both vary with temperature, which should remain constant over the course of the measurement. To use the method to investigate magnetic field effects, permanent magnets are preferred to electromagnets, which have the drawbacks of heating and restricting access.

There are three ways a magnetic field could influence the shape of a pendant droplet.

(1) Maxwell stress
(2) A change in the surface tension of the liquid
(3) Kelvin force in the direction of the field gradient when the field is non-uniform.

The first two effects can be produced by a uniform applied field, the third depends on the field gradient to provide partial magnetic levitation.

Maxwell stress. A uniform field H_0 applied to a droplet of isotropic paramagnetic or diamagnetic liquid, induces a small uniform magnetization $M = \chi H_0$, as indicated in Fig. 8.1c. The 'demagnetizing' effect can usually be neglected in water and dilute solutions of paramagnetic ions on account of their small susceptibility, although this is not the case in strongly paramagnetic ferrofluids with $\chi \sim 1$, which minimize their total energy in the magnetic field by adopting the characteristic spiked surfaces described by Rosensweig. The pressure is not continuous at an interface between magnetized liquid and vacuum, where the magnetic normal traction of Eq. 8.6 appears in a direction perpendicular to the interface [54]. There is also a tangential component S_θ of the Maxwell stress in the Lorentz formulation, and the components as given by Datsyuk and Pavlyniuk [16] when the susceptibility is small are

$$P_r = {}^1\!/_2\mu_0 M^2 \cos^2\theta$$
$$S_\theta = {}^1\!/_2\mu_0 MH \sin 2\theta. \tag{8.7}$$

The tangential component for water is the larger by five orders of magnitude ($1/\chi$). Integrating the component of the normal stress in the **z**-direction of the magnetic field over a hemisphere the stress in the z-direction is $(1/4)\mu_0 M^2$. A consequence of the deformation of the magnetized droplet due to Maxwell stress is an increase in its surface area. The deformation $\delta d/d$ of the in the direction of the field, where d is the droplet diameter is a function of the dimensionless quantity $\mu_0 M^2 d/4\gamma$, where γ is the surface tension of the liquid and for small deformations of a pendant droplet they should be roughly equal. Hence, we can estimate that the deformation of a 4 mm pendant water droplet due to the normal Maxwell stress in 1 T as approximately

4.5×10^{-7}. However, if we take account of the **z**-component of the tangential Maxwell stress $(1/3)\mu_0 MH$, the estimated deformation in 1 T is 4%, or 1% in 0.5T.

A similar result is obtained by relating the Maxwell stress to the magnetic Bond number B_m, the dimensionless ratio of magnetic to surface energy, defined as $B_m = \chi B^2 d/2\mu_0\gamma$. This ratio is 0.2 for a 4 mm droplet of water in 1 T and one or two orders of magnitude greater in solutions of paramagnetic ions. The deformation of the droplet in a uniform applied field is proportional to the square of the field, the susceptibility of the liquid and the size of the droplet itself [21, 22].

The field produced by a single permanent magnet is inherently non-uniform in magnitude and direction. To achieve an approximately-uniform field over a droplet, the magnets must have dimensions that are several times that of the drop. A simple configuration uses two identical rectangular magnet blocks magnetized the same direction, with the drop in the airgap between them. The permanent magnet configurations in Fig. 8.2a and b are assemblies of small magnet cubes designed apply a vertical magnetic field gradient ∇B to a pendant droplet.

Measurements of the influence of magnetic field on the shape of a pendant drop are conveniently made by first observing the drop in zero field in an optical contact analyser, then raising the permanent magnet array so that the drop is immersed in the field and after several minutes of observation lowering the magnets and observing the drop again in zero field. In this way it is possible to correct for any drift. An example of a measurement of water in the nonuniform field produced by the magnet arrays of

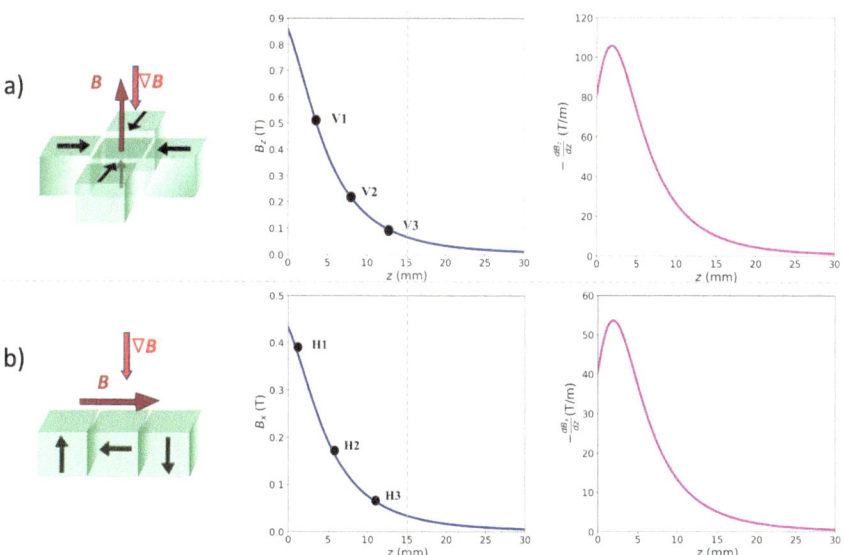

Fig. 8.2 Magnet arrays that produce **a** a vertical field with a vertical gradient and **b** a horizontal field with a vertical gradient. Black arrows show direction of magnetization of the green permanent magnets. The variation of B and $\nabla_z B$ with height above the centre of each magnet array is shown. The measurement positions in the vertical and horizontal fields are noted

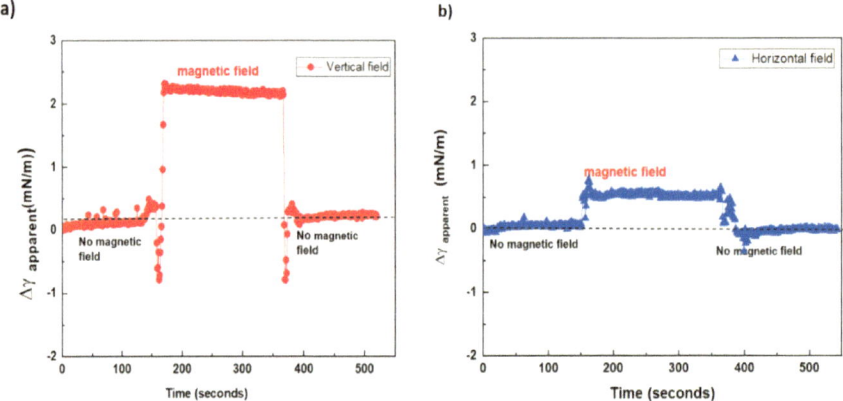

Fig. 8.3 Left. Typical data on a drop of diamagnetic water in no field and in a vertical gradient field. The droplet shape in terms of an 'apparent surface tension' as the magnets are raised to influence the drop, and then lowered again to establish any baseline drift

Fig. 8.2a and b is shown in Fig. 8.3. All three effects listed above are simultaneously present, but the Kelvin force is dominant, and it tends to partially levitate a drop of pure water since $B\nabla_z B$ at the centre is -43 T^{2}m^{-1}, a fraction of that given by Eq. 8.5. The points on the figure correspond to fitted images, with the normal value of g, the acceleration due to gravity, and a change in 'apparent surface tension' γ_{eff} that best fits the shape.

The deformation due to the magnetic field gradient effect is the greatest whenever it is present, but it is eliminated when the field is uniform. The influence of a uniform 450 mT field on γ_{eff} of deionized water is an increase of 0.19 ± 0.21 mNm^{-1}. The error is the standard deviation on the mean of 37 measurements. The contribution of Maxwell stress to the deformation of a pure water droplet is negligible.

8.3 Surface Tension

8.3.1 Static Surface Tension

Returning to the question of surface tension, it is possible to eliminate any direct influence of magnetic forces on a liquid that has exactly zero susceptibility. This is achieved in dilute solutions of the right concentration of a paramagnetic ion to cancel the diamagnetic susceptibility of water, giving a net zero-susceptibility aqueous solution. Any effect of a magnetic field on surface tension should then reflect its influence on the chemical bonding in the water. The zero-susceptibility molar concentration of paramagnetic ions is calculated by setting to zero the sum of the diamagnetic contribution of water (dimensionless susceptibility, $\chi_w = -9 \times 10^{-6}$).

Table 8.2 Calculated molar susceptibility at 295 K, and calculated and measured zero-susceptibility concentrations

	Cu^{2+}	Mn^{2+}	Dy^{3+}
m_{eff}	1.73	5.92	10.65
χ_{mol}	16×10^{-9}	187×10^{-9}	604×10^{-9}
x_0	0.56	0.048	0.015
$x_0{}^{exp}$	0.47	0.051	0.017

and the positive Curie law susceptibility the of $3d$ or $4f$ ions

$$\chi = \mu_0 n g^2 p_{eff}^2 \mu_B^2 / 3 k_B T. \tag{8.8}$$

were n is the is the number of ions per unit volume, g is the Landé g-factor and the effective Bohr magneton number p_{eff} is $(S(S+1))^{1/2}$ for $3d$ ions and $(J(J+1))^{1/2}$ for $4f$ ions where S and J are the spin quantum number and the total angular momentum quantum number, respectively [13]. The effective ionic moment is $p_{eff}\mu_B$ where μ_B is the Bohr magneton. The numerical expression for the susceptibility of a mole of ions is $\chi_{mol} = 1.571 \times 10^{-6} p_{eff}^2/T$. The susceptibility of an aqueous ionic solution of molarity x in the dilute limit is therefore

$$\chi = \chi_w/1000 + 1.571 \, 10^{-6} x p_{eff}^2/T. \tag{8.9}$$

where the first term is the susceptibility of a litre of water. Hence, we can calculate the molar concentration x_0 of any paramagnetic ion where the net susceptibility of the solution is zero. Table 8.2 lists the molar susceptibility of Cu^{2+}, Mn^{2+} and Dy^{3+}, the three ions we will consider here, as well as the molarity $x_0 = -9 \times 10^{-9}/\chi_{mol}$ of the zero-susceptibility solution where χ_{mol} is the molar susceptibility of the ions, given by Eq. 8.6 with $n = N_0$ and using a Landé g-factor of 2 for $3d$ ions and 4/3 for Dy^{3+}. Results for the three ions at $T = 295$ K are shown in Table 8.2. Also included are the experimental values of $x_0{}^{exp}$ measured by SQUID magnetometry, which are a little different.

From measurement of the zero-susceptibility solutions, we deduce that the change is surface tension in a 0.5 T gradient field is -0.40mNm^{-1} for 0.47 M Cu, -0.48mNm^{-1} for 0.051 M Mn and -0.30mNm^{-1} for 0.017 M Dy. Similar results, -0.11mNm^{-1} for 0.47 M Cu, -0.51mNm^{-1} for 0.051 M Mn and -0.50mNm^{-1} for 0.017 M Dy are obtained in a 0.4 T uniform field because the Maxwell stress is then negligible [51]. These are all small changes, barely outside experimental error of zero. A superconducting magnet and a different means of measuring surface tension is needed for a more significant result. Such measurements have been done by Fujimura and Iino who find an increase of just 1.83 mNm^{-1} in 10 T for pure water [26] or 0.09 mNm^{-1} in 0.5T.

8.3.2 Dynamic Surface Tension

Considering the influence of magnetic fields on surface tension in a dynamic context provides an interesting insight into their effects on water. Water has been suggested to have a dynamic surface tension that can be inferred by examining the process of droplet pinch-off, a highly surface tension-driven event [29]. When a drop detaches from an orifice, a thin filament is first formed as the drop pulls away under gravity, which elongates while decreasing in diameter until the filament breaks and the drop detaches. At times close to the filament breaking or pinch-off, the thinning behaviour of an inviscid fluid can be described by a universal scaling law

$$D_{min} = A(g/r)^{1/3}t^{2/3} \qquad (8.10)$$

where D_{min} is the minimum filament diameter, A is a universal prefactor and t is the time from detachment (t-t_0) [37]. Some evidence has been reported that shows viscous effects still have an effect for low viscosity fluids such as water even for filament diameters much greater than the viscous length scale, $l_v = \mu^2/\gamma\rho$, for dynamic viscosity μ [17]. Pinch-off in this scaling regime is independent of gravity as the filament diameters close to pinch-off are much lower than the capillary length $\kappa^{-1} = (\gamma/\rho g)^{1/2}$ (2.72 mm for water) (Fig. 8.4).

Applying a gradient field to a detaching droplet has the effect of modifying the overall body force the fluid filament experiences due to the Kelvin force. This is evident when observing the pinch-off of a fluid such as 0.11M $DyCl_3$ which is more magnetically susceptible than water. Figure 8.5b shows the fluid filament of a 0.11M $DyCl_3$ drop pulled towards a permanent magnet placed to the right of the drop resulting in an angle of tilt of 8°, while a much smaller tilt is observed in water in Fig. 8.5a. The modified body force results in the effective gravity the drop experiences changing by a small amount. However, as the filament diameters close to pinch off are much smaller than the capillary length of 2.72 mm for water, the

Fig. 8.4 Minimum filament diameter of water as a function of dimensionless time t*, in which the time to pinch-off is normalised by the capillary time $t_c = (\rho D_0{}^3/\gamma)^{1/2}$. Data for water in no magnetic field as well as three different gradient fields superpose. Inset: diameters at times close to pinch-off in which the proposed scaling law should apply. No effect of the magnetic field can be observed

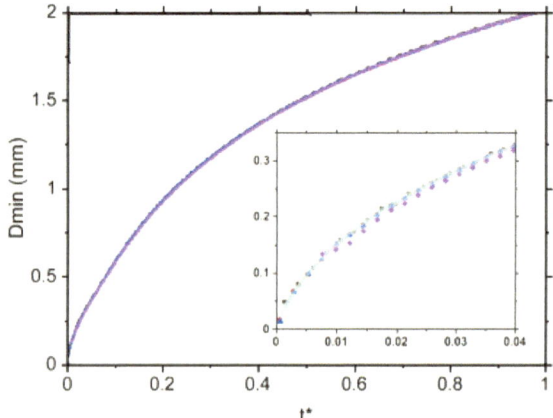

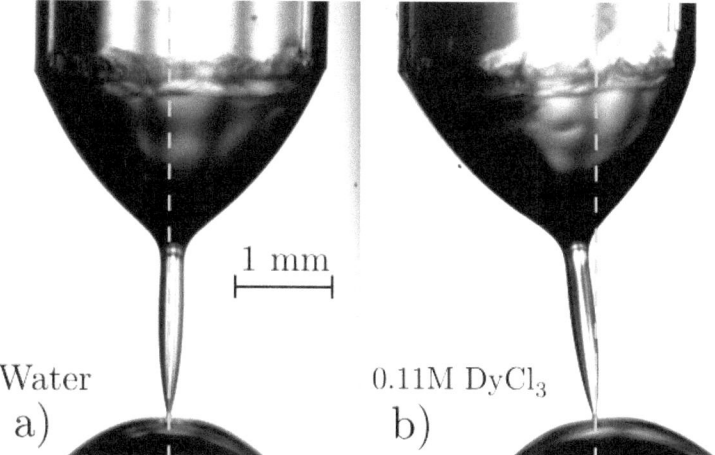

Fig. 8.5 Pinching filaments of **a** water and **b** 0.11M $DyCl_3$ in the presence of a gradient magnetic field due to the presence of a permanent magnet placed to the right of the filament in each image. A greater degree of tilt is observed for the more highly susceptible $DyCl_3$

small change in the effective gravitational force due to the presence of the gradient field does not affect the pinch-off dynamics, even in the case of 0.11M $DyCl_3$ after correcting the image for tilt. Using high speed photography, we have shown this to be case as no measurable effect on the pinching dynamics by a gradient field was observed [51].

8.4 Evaporation Rate

A possible influence of magnetic field on evaporation rate of water has attracted the attention of experimentalists in recent years, and a body of often-contradictory experimental results has accumulated. Some of the reports of persistent magnetic field enhancement of water evaporation rate [11, 24, 28, 40, 46, 55] been reviewed by Chibowski and Szcses [10]. Here we address the questions of if and why a static or intermittent magnetic field can change the evaporation rate? Whenever possible, control experiments where relevant environmental factors such as airflow, relative humidity and temperature are carefully controlled, or continuously monitored as two samples, one in magnetic field and the other a no-field reference are simultaneously subject to otherwise identical, but possibly time-varying ambient conditions.

Various types of experiments have been carried out: First, are those where the weight of an open vessel of water is measured while it is exposed to a static field B and compared to a no-field reference [2, 4, 11, 40, 55]. These sometimes involve interrupting the experiment at different times to weigh the vessel and the remaining water. A better approach is to monitor an in-field sample and a no-field control simultaneously. An alternative protocol is to measure the loss of weight by evaporation in zero field at different times after a single brief exposure to a static field [32], and compared with an unexposed sample in the same conditions. Here, a persistent memory effect of magnetic field exposure on the water is sought. The field is commonly produced by a permanent magnet, rather than an electromagnet, to avoid any heating but the field is inevitably nonuniform to some extent in magnitude and direction over the evaporating water surface unless the surface area is very much smaller than the airgap of the magnet. Fields used range from tens to hundreds of millitesla.

Second are those types of experiment where the water is exposed to a nonuniform field, by continuous flow at a velocity $\gtrsim 1$ cms^{-1} around a circuit where a permanent magnet surrounds a section of the pipe [2, 23, 58, 60]. The experimental setup resembles that used for magnetic treatment of hard water to control limescale, except that the water is usually circulated not just once but repeatedly through the inhomogeneous magnetic field, which is equivalent to periodically exposing the water to ~ 20 ms magnetic field pulses at frequency of about 1 Hz. After a fixed circulation period with one or many pulses, the magnetically-treated water is removed and its rate of evaporation is tracked by weighing. Remarkably, both types of experiments give qualitatively similar results—a modest increase of evaporation rate is usually reported. They have been combined in a two-stage experiment to increase the effect [62].

There is little consistency in the experimental protocols adopted, and results depend on humidity, temperature, magnetic field and time in different ways. Reported increases in evaporation rate associated with the magnetic field range from a few percent to more than 30% [4, 8, 11, 23, 32, 40, 55]. The magnetic field was thought to modify somehow the network of molecular hydrogen bonding in water, but there is no agreed explanation of how this might occur. It must be a subtle phenomenon—the direct decrease in energy of $\frac{1}{2}\chi_{mol}B^2/\mu_0$ of water with molar susceptibility $\chi_{mol} = -1.6 \; 10^{-10}$ [13] in a field of 1 T is—64 μJ/mol, seven orders of magnitude less than the energy of a hydrogen bond in water [57].

The third type of experiment less controversial. Here water was exposed to an intense magnetic field of order 10 T in a superconducting solenoid, with a large horizontal gradient $\nabla_x B$ so that the product $B\nabla_x B$ is 320 T^{2}m^{-1}. The explanation here [46] lies in the paramagnetic susceptibility of atmospheric oxygen, which is displaced by the evaporating water vapor. The magnetic forces are 17% of the buoyancy forces on air, which leads to a field-induced modification of gaseous convection and hence the water evaporation rate.

8.4.1 Droplets

There are reports in the literature of the evaporation of water in sessile and pendant droplets. Notable are interferometric measurements based on Newtons rings from a sessile droplet by Verma and Singh [59] who reported an increase of evaporation rate of 28% in the presence of a 100 mT magnetic field. This method is rapid and accurate with a precision of 5 nm for the droplet height.

Evaporation rates of water can be modified by dissolved ions that have the propensity to make (kosmotropes) or break (chaotropes) the hydrogen-bonded structure of liquid water and thereby reduce or increase the evaporation rate compared with pure water. This influence is reflected in the Hofmeister series of dissolved cations or anions [53]. The evaporation rate of droplets on hydrophilic surfaces increases with salt content, and Marangoni convection of the solute, which is due to different evaporation rates at the centre and edge of the droplet that leads to a radial thermal and surface-tension gradients, is the dominant cause of advective flows in the solutions [36]. Flow in the liquid entrains a convective flow in the adjacent air that modifies the evaporation rate.

Similar flows in pendant droplets are modified by uniform magnetic fields in the 100 mT range acting on conducting ~ 0.1 M solutions of $FeCl_3$ $CoCl_2$ and $NiSO_4$, with significant enhancement of the evaporation rate [35]. These are essentially magnetohydrodynamic damping effects the Lorentz force

$$F_L = \sigma (v \times B) \times B \qquad (8.11)$$

acting on the Marangoni flow in a liquid with conductivity σ, an explanation that does not depend on the magnetism of the dissolved cation. No increase of surface tension was found by Jaiswal et al.

8.4.2 Confined Water and Aqueous Solutions

Our own work has focussed on simultaneous measurements of twin samples in the same Perspex enclosure to restrict airflow, one exposed to a quasi-uniform 300–500 mT field, the other a no-field control. Temperature and relative humidity are monitored automatically for periods of up to 60 h. Evaporation of water or aqueous solutions was investigated either from open beakers or from tiny drops centred in microchannels. The twin magnetic setups are illustrated in Figs. 8.6 and 8.12.

8.4.2.1 Open Beakers

Two 100 ml beakers are half-filled with water or aqueous solution and placed on separate balances in the Perspex enclosure. One is surrounded by a large Halbach ring magnet producing a quasi-uniform magnetic field over the liquid surface close to 500 mT, with a field gradient reaching 3 Tmm^{-1} at the edge. (Fig. 8.6). Evaporation is proportional to the surface area, and the weights of the two beakers are monitored for periods from 1 to 60 h. Some short-time runs were also made using a single beaker on a normal laboratory balance with doors shut to exclude draughts [52] for reasons that will become clear.

Deionized water. Most work, including 36 runs of 16 or 60 h, was done with Millipore deionized water. In every case, evaporation was greater in the presence of the magnetic field. The average enhancement was $12 \pm 7\%$ [52] (Fig. 8.6).

The evaporation rate averaged over all runs for which RH $= 0.73 \pm 0.06$ is 0.0297 ± 0.0081 kgm^{-2} h^{-1} without field and 0.0331 ± 0.0088 kgm^{-2} h^{-1} in the 500 mT

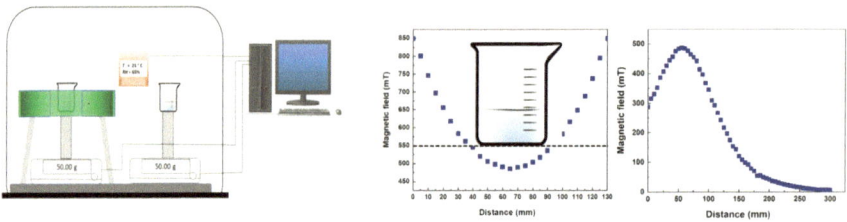

Fig. 8.6 Experimental arrangement where in-field and no-field evaporation of water is monitored continuously by measuring the time-dependence the mass of water in beakers. Magnetic field profiles are shown on the right

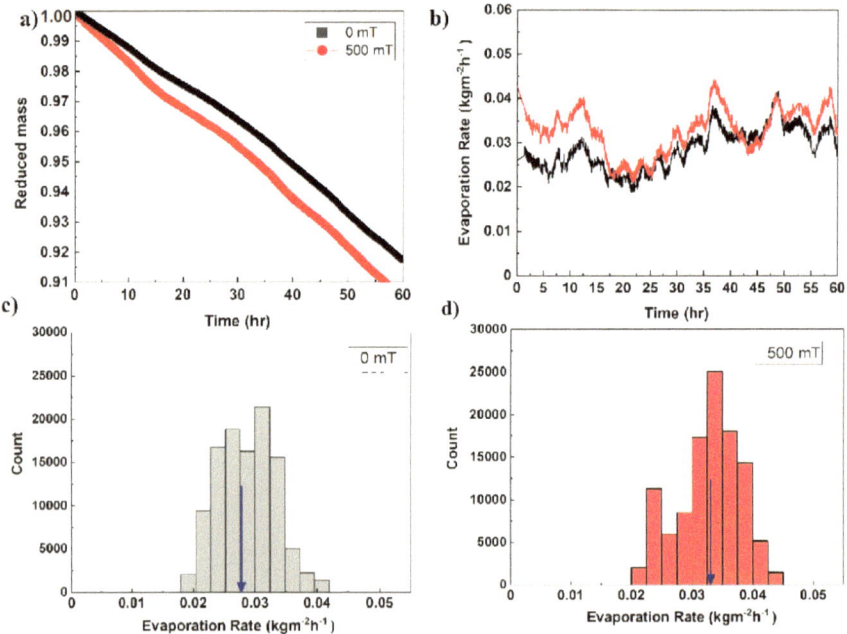

Fig. 8.7 Extended 60 h run of evaporation of water versus time for a 50 mL sample of water in 500 mT (red) and the no-field reference (black, grey). **a** relative weight loss by evaporation **b** evaporation rates versus time and **c** histograms of the evaporation rates, where the average values are marked by blue arrows [52]

field with the field gradient. The rate often fluctuated by 10% or more In the course of the experiments because of slow variations of in ambient temperature and relative humidity. An extended, 60 h run is analysed in Fig. 8.7. Fluctuations due to changes of the ambient temperature and humidity, as well as local variations of evaporation rate in the beakers lead to the fluctuations during the run illustrated in Fig. 8.7b, but it should be noted that at almost any instant, the evaporation rate of the sample in the magnetic field is greater than that of the reference. The net weight loss in the magnetic field in this case was 14% greater than that of the reference at the end of the run.

The evaporation rate varies non-linearly with temperature T which determines the capacity c_a of dry air to absorb water vapor plotted in Fig. 8.8a; the variation around room temperature is 6% per K. It also varies with relative humidity RH. Although we did not control it, the value was often in the range 60–70%. Sometimes the rate was quite steady over a 16-h period, but on other occasions it changed by 5–20%. Evaporation rates of water from half-filled 100 mL beakers versus $(100\!—\!RH)$ with RH in % are plotted in Fig. 8.8b. The data are taken at different times of the year and corrected to 23 °C. A star marks the average value of g obtained from the 36 extended runs where the average humidity was 72.8%. The data suggest that the evaporation rate is proportional to $(100\!—\!RH)$ and they extrapolate to 0.115 (10) kgm^{-2} h^{-1} for dry air.

An empirical formula for the evaporation rate of water in open air, used by engineers is [25]

$$g = \Theta c_a(1\!-\!RH) \tag{8.12}$$

The prefactor Θ is numerically equal to $(25 + 17v)$ kgm^{-2} h^{-1}, where v is the speed of the surface airflow in ms^{-1}. In still, dry air at 23 °C, the predicted value of

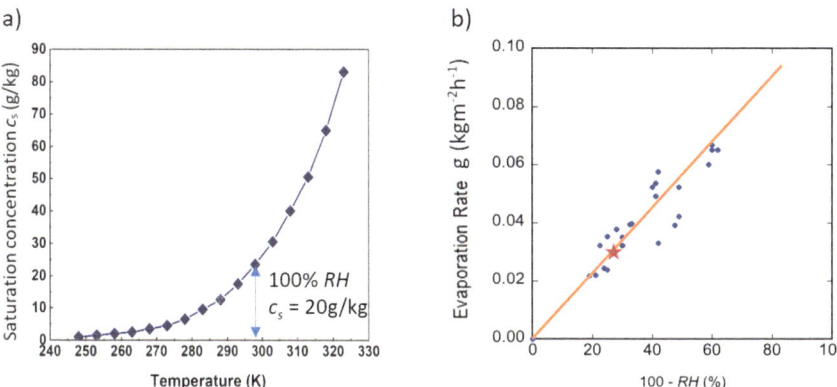

a)

b)

Fig. 8.8 a Variation of c_a the capacity of dry air to absorb water vapor in g/kg as a function of temperature [25]. **b** Variation of evaporation rate g with relative humidity RH, measured in the beaker experiments [50]

g at 50% *RH* is 0.49 kgm^{-2} h^{-1}. The water in the beaker evaporates 45 times more slowly, Fig 8.8b, extrapolated to 100%.

6 M urea solution. Next, we consider a different liquid where 26.5 w/w% of the water has been replaced by urea, a non-volatile liquid that has no appreciable vapor pressure and is expected to disrupt the hydrogen bonding when in solution in the host water. Here we find that the evaporation rate is 0.0422(23) kgm^{-2} h^{-1} at a relative humidity of 49%, which extrapolates to 0.0867 kgm^{-2} h^{-1} in dry air. This is appreciably less than the value for pure water. In urea solution, the magnetic field decreases the rate of evaporation significantly, by 28 ± 6% to 0.330 (41) kgm^{-2} h^{-1} [52].

Salt solutions. Results obtained on ionic salt solutions reveal a rich variety of behaviour both as regards dry evaporation rate and the influence of magnetic field. For example, the evaporation rate of water from lithium, sodium and potassium chloride solutions decrease monotonically with increasing molarity, and it is 50% lower in 4 M NaCl compared to pure water (Fig. 8.9). The effect of magnetic field is to increase the evaporation rate for most concentrations around 1 M. The evaporation rate of a sessile droplet of water in open air is 0.91 kgm^{-2} h^{-1}, and it is not influenced by a magnetic field. However, when the droplet is surrounded by a quartz cuvette partially open at the base so that the droplet evaporates into its own vapor, Fig. 8.10b, the rate is reduced by a factor 20 but errors are large, as seen in Fig. 8.10c.

Time dependence

At this point another experimental result helps to cast light on what is controlling the evaporation of water from beakers. Water is simply poured into a 100 mL beaker, which is placed on a chemical balance with the doors shut to exclude drafts while weight loss is monitored for the first two hours. The beaker again is roughly half full. No magnetic field is applied. The initial transient is quite different from the steady rate reached after about an hour. (This is a reason for using long runs used to measure evaporation rates). Results for deionized water are shown in Fig. 8.11, where data were fitted to the equation

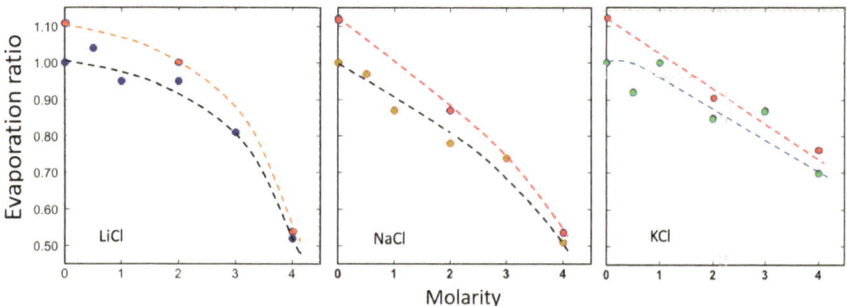

Fig. 8.9 Evaporation rates of water in beakers from salt solutions of different molarity. The trends with and without an applied field are indicated by red and black dashed lines, respectively

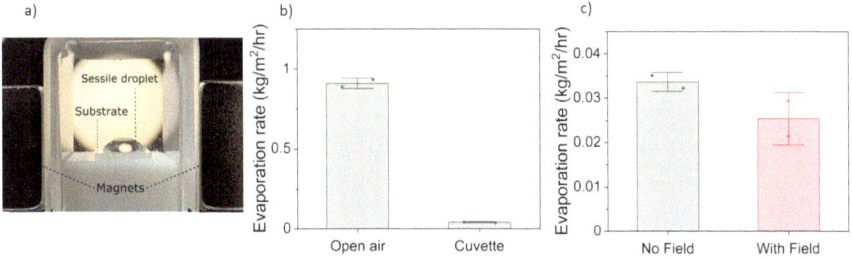

Fig. 8.10 a A sessile droplet of water evaporating in an inverted cuvette, **b** Effect of cuvette on the evaporation rate and **c** Comparison of the evaporation rates of water and 1 M NaCl in the cuvette

$$m(t) = m(0) - (1 - e^{-r/t})(g_0 - gt) \tag{8.13}$$

The mass of water at the start of an experiment is $m(0)$, τ is the decay time of the initial transient, and g_0 and g are the initial and steady state evaporation rates in kgs^{-1}.

The average of eight one-hour runs for water gave an initial evaporation rate $g_0 = 0.134 \pm 0.030\,\mathrm{kgm}^{-2}\,\mathrm{h}^{-1}$ and an average τ of 14 ± 3 min. The long-time evaporation rate was only half as great $0.068 \pm 0.010\,\mathrm{kgm}^{-2}\,\mathrm{h}^{-1}$, Table 8.3.

These key difference between the early and later stages of evaporation is that initially water is evaporating into ambient air whereas later it is evaporating into air

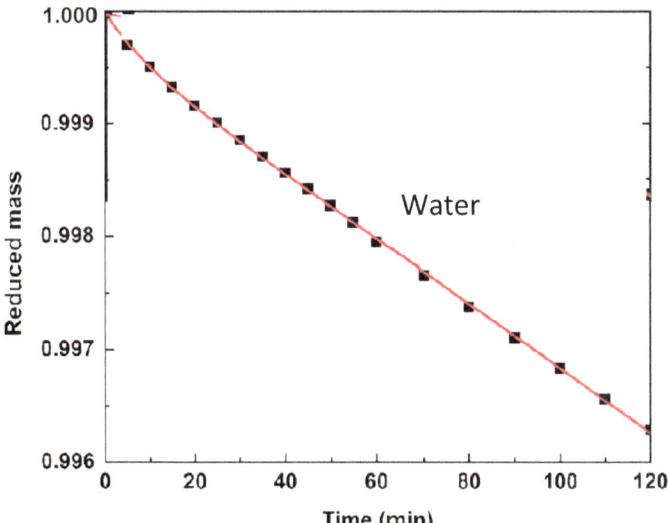

Fig. 8.11 Short-term evaporation of Millipore water in stagnant conditions, showing the steady-state and initial transient regimes, measured in a chemical balance. The initial transient, which decays exponentially with a time constant of 9 min is followed by uniform steady-state evaporation after about 30 min

in the beaker where most the water vapor originally in ambient air has been replaced by vapor from the evaporating liquid. How they differ is discussed in §6. When deionized water is replaced by heavy water, the evaporation rate is less than half as great $(0.013\ \mathrm{kgm}^{-2}\mathrm{h}^{-1})$ and no initial transient is observed (Table 8.3).

From the typical evaporation rates, which depend on both temperature and relative humidity, and a density of air of $1.2\ \mathrm{kgm}^3$, the time taken to evaporate enough fresh water vapor to create a relative humidity of 50% in the empty half of the beaker of order 20 min. The average time taken for a water molecule escaping from the surface with a root mean square velocity deduced from equipartition of energy of $630\ \mathrm{ms}^{-1}$ to diffuse out of the beaker is of order a minute or two. The evaporation at the surface is controlled by a mass balance of freshly escaping water vapor and recondensing molecules from region a within a few mean free paths of the surface, where the freshly evaporated molecules become the majority after a time τ.

Table 8.3 Initial and steady-state water evaporation rates and initial decay time in no field

Initial evaporation rate g_0 $(\mathrm{kgm}^{-2}\mathrm{h}^{-1})$	Decay time τ (minutes)	Steady evaporation rate g $(\mathrm{kgm}^{-2}\mathrm{h}^{-1})$	T (°C)	RH (%)
0.134	14	0.068	24	60

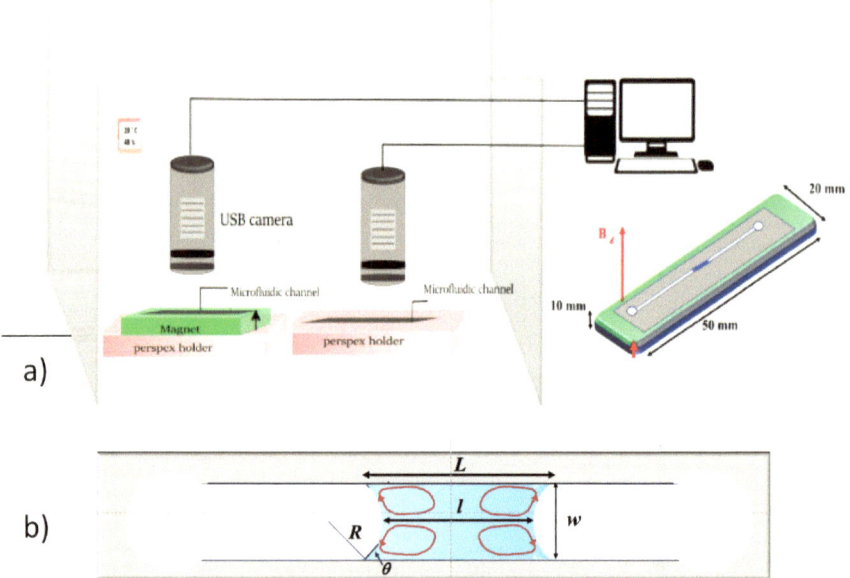

Fig. 8.12 **a** Experimental arrangement where in-field and no-field evaporation of water is monitored continuously by measuring the volume of water in two PMMA microchannels; **b** Dimensions of the evaporating water sample at the centre of the channel, with a schematic illustration of Marangoni flow

When evaporation is measured from aqueous solutions as a function of time it is important to establish thermal equilibrium at ambient temperature before beginning measurements in the balance. Some salts like LiCl are exothermic when mixed with water, but urea is endothermic. In these cases any initial transient may be masked by the effects of temperature changes on the evaporation rate.

8.4.2.2 Water in Microfluidic Channels

Following from the idea that confinement of the fresh water vapor may be necessary to observe a magnetic field effect on the evaporation rate, we have performed experiments in microchannels analogous to those described in Sect. 8.4.2.1 [50]. The channels are made from three layers of poly(methyl methacrylate) (PMMA) assembled by thermo-lamination after cutting a 1 mm wide channel, 54 mm long in the 0.38 mm thick middle layer with a CO_2 laser. The experimental arrangement is illustrated in Fig. 8.12. No magnetic field is applied to water in one channel while water in the other is exposed to a 300 mT field perpendicular to the channel, produced by rectangular $50 \times 20 \times 10$ mm^3 Nd-Fe-B magnets. Evaporation is monitored by simultaneously imaging 0.4 µL droplets of Millipore water positioned at the centres of the two 50 mm channels with PCE800mm USB cameras The ends of each microchannel are open to ambient air, and the evolution of the shapes of the drops are recorded as they shrink down to a membrane after several hours and rupture shortly afterwards. The twin setup was enclosed in a perspex box and the ambient temperature (26 °C) and relative humidity RH (42%) in the laboratory were controlled throughout.

Menisci appear at the edges of the drops with contact angles θ of about 30°, as seen in Fig. 8.12. Water normally exhibits a contact angle $< 90°$ on PMMA [63], but the menisci form because of the periodic variation of the channel width due to cutting with the CO_2 laser. The variation is 10% of the 1 mm channel width with a period of 300 µm. Just like a magnetic domain wall, the water seeks to minimize its surface area and form a circular meniscus because of surface tension. As the water evaporates, the meniscus jumps to a neighbouring pinning point in a stick–slip process. The volume of water in the confined drop is deduced from its approximately symmetric shape parameterized by L, the average of the contact lengths at the two edges and l, the distance between the centers of the two menisci, as shown in Fig. 8.12b. The volume V of liquid, calculated from L and l knowing the width w and the depth d of the channel, is

$$V \approx wd\,(L - 2w\theta/3) \tag{8.14}$$

where contact angle θ is $\tan^{-1}\left[(2R\text{-}L + l)/w\right]$ [50].

Some results are illustrated in Fig. 8.13; Figure 8.13a captures the last moments of a sample of water that shrinks to a membrane, and then ruptures, while Fig. 8.13b shows a typical sample that takes almost four hours to evaporate. Sessile droplets of recondensed water vapor can be seen growing in the channel beyond the menisci, which indicate that the air in the channel is saturated with water vapor. The areas of the menisci from which evaporation is occurring remains roughly constant and equal to $2wd$.

Data for two representative runs are shown in Fig. 8.14. Initial values of L and l are normalized to 1. The runs, like those in the beakers, showed significant variability. The average evaporation rate from the control channel over ten runs was 0.13 ± 0.03 kgm^{-2} h^{-1}, which is four times that in the beakers. The magnetic field enhancement was much greater, ranging up to 140%, with an average of $61 \pm 42\%$. The stable sessile droplets of water with diameters of 30–150 µm growing slowly in the channel throughout a run were observed in most of the measurements when a magnetic field was applied. They appear at distances greater than about 100 µm from the menisci in Fig. 8.2b, which means that the air in the microchannel at 299 K is be saturated with water vapor, and they may be a result of the enhanced evaporation rates shown in Fig. 8.14.

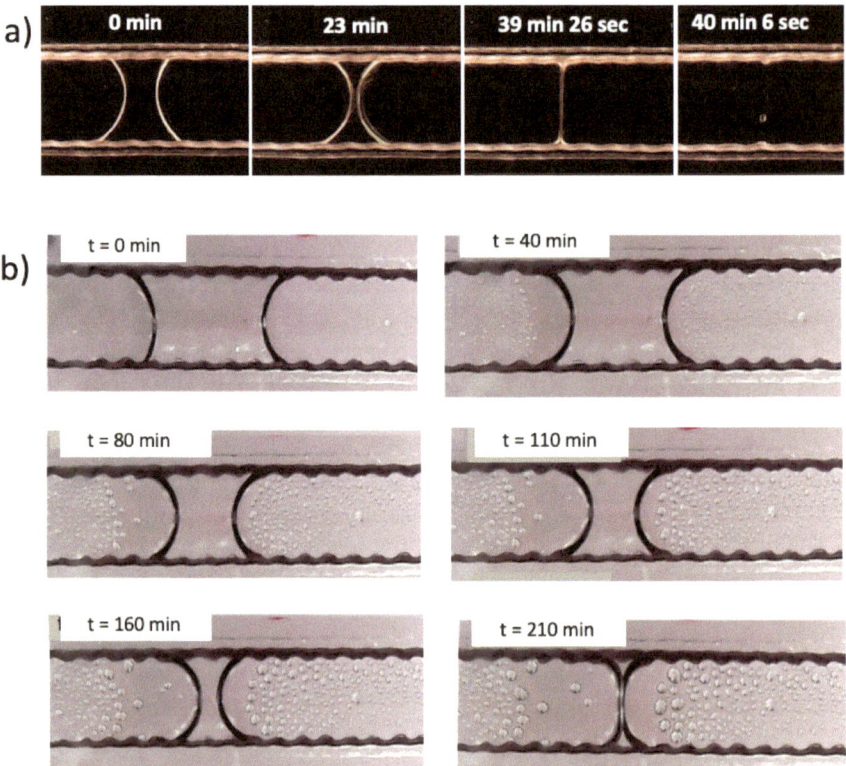

Fig. 8.13 a Evolution of 0.12 µL drop of water in a microfluidic channel. After 39 min it has shrunk to membrane that ruptures 30 s later. **b** A typical example of evaporation of a 0.5 µL drop. Sessile droplets of recondensed water vapor grow in the channel beyond the menisci, indicating that relative humidity there is 100% [50]

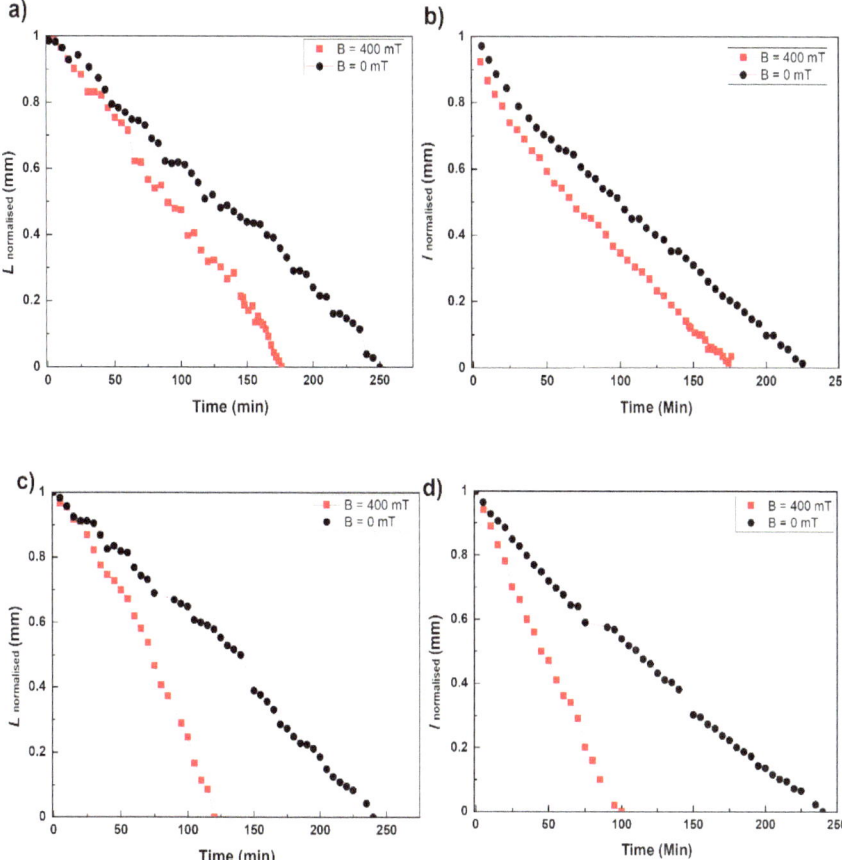

Fig. 8.14 Variation of L and l, normalized to the values at $t = 0$, during two evaporation experiments; **a** and **b** are for experiment 1. **c** and **d** are for experiment 2. Data in black circles are for the no-field channel and data in red squares are for the channel in a 300 mT magnetic field

The flow dynamics of water in the microchannel and in the beaker must be quite different. The flow regime in the two cases may be characterized by the inverse Bond number, the dimensionless ratio $- (\partial\gamma/\partial T)/\beta\rho g d^2$ of Marangoni to Rayleigh numbers, which reflects the relative importance of the surface tension and gravitational forces that influence the flow. Here, $\beta = 2.1 \times 10^{-4}\,\mathrm{K}^{-1}$ is the thermal expansion coefficient of water and d is the depth. The ratio is 316,000 for the microchannels and 90 for the beakers. Surface tension dominates flow in the channel. The faster evaporation rate of water in the microchannel compared with the beaker is attributed to Marangoni convection. Thermocapillary flow with the vortex pattern sketched in Fig. 8.12b, like that observed at a meniscus in a capillary [6, 7, 20], will increase evaporation in the microchannel. Magnetic field is known to influence the flow in droplets of ferrofluids and conducting fluids [19, 35] (Eq. 8.9), but the deionized water we use is neither conducting nor magnetic. We need a different explanation.

The vapor close to the meniscus is composed of freshly evaporated molecules, which need high kinetic energy and well-timed making and breaking of at least three hydrogen bonds at the interface in to break loose [45]. The isomer ratio of such water vapor may differ from that in liquid water and its effective temperature will be higher than ambient in the channel, allowing it to evaporate without immediately recondensing. The Knudsen layer where there is a temperature gradient normal to the water surface [34] should be wide enough to allow evaporation to proceed into the channel at the ambient temperature of 26 °C.

8.5 An Explanation

8.5.1 Isomers of Water

We saw in the discussion of Fig. 8.11 that water vapor freshly evaporated at a liquid surface seems to be different from that in equilibrium in the atmosphere. Not all molecules of H_2O are the same. Water exists as one of two nuclear isomers, according to whether the two proton spins are aligned parallel with net nuclear spin $I = 1$, or antiparallel with net nuclear spin $I = 0$, as shown in Fig. 8.15. Molecules with a nuclear spin triplet or singlet are known (somewhat confusingly) as *ortho* or *para* water respectively and every molecule is in one state or the other. To conserve angular momentum, the *ortho* isomer has an angular momentum of $\hbar$ in its ground state and the *para* isomer, which has none, is lower in energy by 2.95 meV (34.3 K). There are three possible spin substates with $I = 1$, namely $I_z = 1, 0$ and -1, but only one, $I_z = 0$, with $I = 0$. The spin substates are respectively symmetric $|\uparrow\uparrow\rangle, |\uparrow\downarrow\rangle + |\downarrow\uparrow\rangle, |\downarrow\downarrow\rangle$ and antisymmetric $|\uparrow\downarrow\rangle - |\downarrow\uparrow\rangle$. The water vapor in equilibrium at ambient temperature is expected to exhibit an *ortho:para* ratio of 3:1, and this has been confirmed by terahertz spectroscopy of the vibrational energy levels [44] in the 0–3 THz range. However, the ratio in liquid water, which will be influenced by hydrogen bonding, is thought to be close to 1:1 [47], and the ratio in the escaping water vapor, which has to be involved in a three-body collision in order to break free [45] will be different again.

Unlike the corresponding isomers of hydrogen gas, H_2, which differ in energy by 15 meV in their ground states and can be separated in the gas phase just above the boiling point and then used for dynamic nuclear polarization to enhance the sensitivity of nmr, it has proved very difficult to separate ortho and para water by physical methods. Separation rates of 1 pL per day have been achieved in molecular beams, but the separated liquid isomers have half-lives of about an hour at ambient temperature [33] and 3:1 equilibrium is soon re-established. However, the molecular isomers in the vapor phase are much longer-lived and equilibrium can take weeks to establish [9].

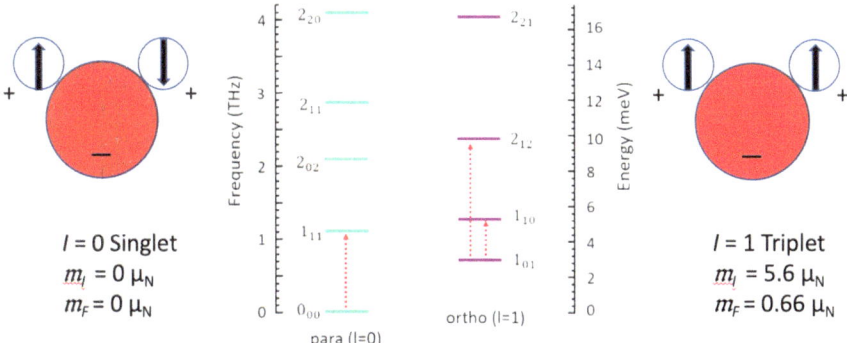

Fig. 8.15 *Para* (left) and *ortho* (right) water molecules. The spin and orbital moments of each are listed and the vibrational energy levels are shown in the centre

Nuclear singlet–triplet transitions are strongly forbidden [1], so *ortho* and *para* water vapor may be regarded as separate molecular species—two quasi-independent gasses. This hypothesis allows us to rationalize the observed magnitudes of the evaporation rates with and without a magnetic field in terms of the ortho:para ratio $f_v^\circ{:}f_v^p$ of the escaping water vapor.

8.5.2   Analysis

Data analysis is based on the hypothesis that *ortho* and *para* water vapor behave as independent gasses on the timescale of the experiments. The ratio in freshly-evaporated liquid f_L is evaluated from the data, assuming the re-condensation rate is constant and isomer-independent, and the effect of the magnetic field is to modify the isomeric ratio in the vapor. The main difference between the beaker, microchannel or confined sessile droplet and water in open air is that the water will be evaporating into its own vapor in a steady state, with an isomer ratio $f^\circ{}_L{:}f^p{}_L$ whereas in open air water is evaporating into ambient water vapor with an isomer ratio $f^\circ{}_V{:} f^p{}_V$ of 3:1. Water evaporating in a sheltered space the originally filled with ambient air, will gradually replace the ambient air by air with a the isomer ratio $f^\circ{}_L{:}f^p{}_L$ of freshly evaporated water. If the space is not sheltered but open to air currents, the ratio will remain 3:1, and magnetic field has no influence on the evaporation rate. This is the case, for example, for unshielded sessile droplets.

The hypothesis is expressed as the ansatz that the evaporation rate is proportional to the dimensionless quantity

$$g_0 = \left[f_L^\circ \left(1 - f_V^\circ\right) + p_L^p \left(1 - f_V^p\right)\right] \tag{8.15}$$

The plot of g_0 as a function of $f^o{}_L = (1 - f^p{}_L)$ in Fig. 8.16 illustrates the evaporation rate for three different conditions of the surrounding vapor. The grey line $f^o{}_V = 0.75$ represents the 3:1 equilibrium ratio in open air. The green parabola represents the situation where the water is evaporating into its own vapor in a confined space. The horizontal red line is for the case where the ratio is 1:1. The horizontal dashed line represents the recombination rate c, which is supposed to be a constant independent of $f^o{}_L$; the net evaporation is measured as the height of the grey, red or green line above the dash line. The net value of g_0 is multiplied by a factor Θ_{exp} to give the actual mass evaporation rate in $kg m^{-2} h^{-1}$.

The effect of water vapor escaping from a from a liquid and accumulating in half-filled beaker is often to reduce the initial rate from the gray line to that given by the green parabola as time goes on. This means that $f^o{}_L$ for water lies in zone A_1 or A_2. There is no effect for $f^o{}_L = 0.5$ or 0.75. When the initial rate increases with time, as it does for 6 M urea or 1 M NaCl, $f^o{}_L$ lies in Zone B. The values of $f^o{}_L$ and c can be determined more precisely from the no-field to in-field evaporation ratio, which is < 1 in almost all cases.

Magnetic field can alter the isomer ratio in the vapor, and hence the evaporation rate, in two ways [52]. One is via Larmor precession at 43 MHz/T, which tends to dephase the two protons when a field gradient is present, making $f^o{}_V = f^p{}_V$ (red line), The other is via Lorentz stress on the electric charge dipole, which tends to increase angular momentum of the H_2O. In Zone A_1 the field effect is always positive. In Zone B it may be negative.

Applying these ideas to pure water we find $f^o{}_L = 39 \pm 1\%$ [52]. In the 2 M and 4 M Li and K solutions the value increases to 0.46 for Li and 0.44 for K, assuming

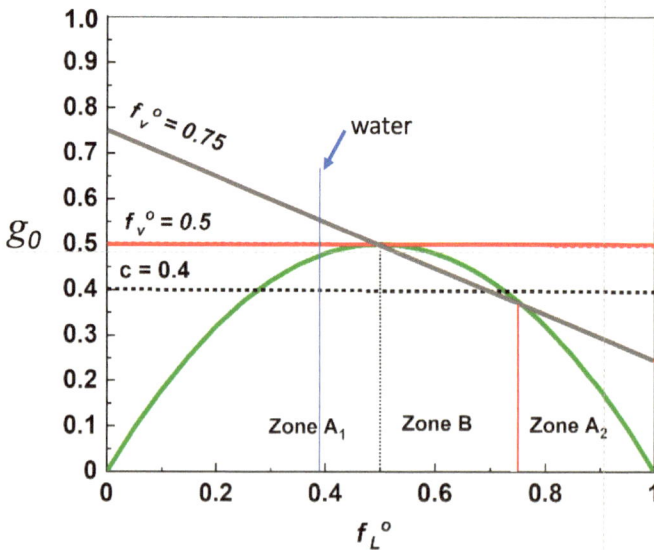

Fig. 8.16 Evaporation diagram based on Eq. 8.15. See text for explanation

c remains 0.4. Na is different, 2 M is similar to water and for 4 M the ortho ratio has increased to 0.46, but for 6 M urea has shifted into Zone B, with a negative field effect and $f°_L \approx 0.6$.

8.6 Crystallization from Solution

Finally, we summarize some information on a related topic—how the crystallization of ionic crystals from supersaturated aqueous solution can be influenced by a static magnetic field, either to promote single-crystal growth or to selectively inhibit the growth of an unwanted crystal polymorph in favour of another. Explanations of both effects involve proton dimers formed during crystal growth.

8.6.1 Ionic Crystals

The effect of a magnetic field on the crystallization of a selection of carbonates phosphates, sulphates and oxalates of calcium, magnesium or zinc, as well as the divalent magnetic ions Mn, Fe or Co were studied by Lundager Madsen [41–43]. He found that when these sparingly-soluble salts are crystallized from solution in a magnetic field of 270 mT, the rates of nucleation and crystal growth are enhanced in the diamagnetic salts of weak acids but there is no effect at high pH, or when the cations are magnetic. A key observation was that the field had no effect when heavy water is used as the solvent. In D_2O, hydrogen is replaced by deuterium, where the nucleus is composed of a proton and a neutron. Both have spin ½ and they couple to give I = 0 making the hydrogen a boson rather than a fermion. The growth of nuclei in solution is assumed to involve the creation of doubly protonated anions on the growing surface, which is possible with no energy penalty for D_2 or H_2 in an I = 0 singlet *para* state, but the triply degenerate I = 1 *ortho* state has a ground state with orbital angular momentum and corresponding vibrational energy that reduces the rate of the process. Nuclear spin relaxation for H_2 tends to equalize the *ortho* and *para* populations and thereby increase the growth rate [43]. There is no such effect for D_2.

8.6.2 Magnetic Water Treatment

Treatment of hard water to avoid precipitation of hard limescale on heated surfaces in domestic and commercial water heaters, boilers and heat exchangers has a long history of mixed practical success and widespread scientific scepticism [3]. The basic claim that is amenable to investigation is that by passing hard water once through a nonuniform magnetic field, typically generated by an array of permanent magnets, it is possible to influence the subsequent precipitation of calcium carbonate when the water is heated. Aragonite tends to precipitate rather than calcite, and the carbonate then has an acicular morphology that does not form hard scale [15, 30, 31, 38, 39]. The water somehow retains a memory of its magnetic exposure that persists for days. The dilemma of reconciling these observations with the rapidly-changing molecular structure of liquid water was lifted after it was discovered that the calcium carbonate dissolved in water was not entirely in the form of separate Ca^{2+} cations and HCO_3 anions, but partly in the form of amorphous polymeric nanoscale prenucleation clusters [27, 49], subsequently named DOLLOPS (Dynamically Ordered Liquid-Like Oxyanion Polymers) [18]. This has led to a new theory of nucleation [27] and also an idea of how magnetic water treatment may work [14]. The requirement is a durable modification of the pre-nucleation clusters during the fleeting exposure of water to the magnetic field. This could be provided by the magnetic field gradient acting on a layer of bicarbonate anions at the surface of the DOLLOP. Adding a Ca^{2+} ion to grow the cluster displaces a proton to form a hydrogen dimer in H_2CO_3 or its stable product $H_2O + CO_2$. Dimerization of the protons is a way to achieve a long-lived modification of the cluster by scrambling the proton singlet and triplet states as their spins are dephased in the magnetic field gradient, where their Larmor precession frequencies ($42.6\,MHzT^{-1}$) are slightly different. The field gradient produced by the magnet arrays is comparable to that needed to dephase the protons by π in the time they take to pass through the inhomogeneous magnetic field [14].

8.7 Conclusions

Magnetic fields have little direct effect on liquid water. Because of its weak diamagnetic susceptibility, exceptionally large magnetic fields and magnetic field gradients are required to produce effects such a magnetic levitation. Compensated solutions of paramagnetic ions with zero susceptibility can be used to avoid extraneous effects of the Kelvin force density on properties such as surface tension. Changes in surface tension measured in pendant droplets are of order -1% per tesla.

Much larger magnetic effects on the evaporation rate of pure water or aqueous solutions are found when the water is evaporating into its own vapor in a confined space sheltered, from external airflow. Increases of 10–100% are observed for deionized water, depending on confinement, but for aqueous solutions the evaporation rate may increase (many salt solutions) or decrease (6 M urea, 1 M NaCl) in the magnetic field. The sign of the field effect on evaporation appears to be associated with the sign of an initial transient in the evaporation rate observed in no field.

The remarkable magnetic field effects on the evaporation rate have been discussed in terms of a two-vapor model, where the two nuclear isomers of water behave as independent gasses. The *ortho:para*, triplet to singlet ratio is 3:1 in equilibrium in ambient air, but in freshly-evaporated water vapor it may be quite different. From the magnitude and sign of the magnetic field enhancement and the shape of the zero-field transient evaporation, it is possible to fix both the isomeric ratio of fresh vapor and the recombination rate to within a few percent. The ratio is 39:61 in Millipore water. The effect of a magnetic field is to modify this ratio in the vapor and thereby modify the rate of evaporation from the liquid surface. Two mechanisms are proposed, one is Larmour precession of the two protons on a single water molecule. A magnetic field gradient will dephase their precession and tend to equalize the two populations. Lorentz torque could augment the angular momentum.

Magnetic field effects on crystal growth from saturated aqueous solutions and magnetic water treatment are similarly related to the nuclear spin of proton dimers. We have not observed magnetic memory effects in pure water.

Surprisingly, the average evaporation rate of water in a microfluidic channel where the relative humidity exceeds 100% is notably greater than it is in an open beaker where the vapor is unsaturated. This is likely due thermocapillary flow in the channel. Future work should aim to visualize thermocapilliary flow in microchannels by particle image velocimetry in a field; evidence of a Lorentz force would reveal any related charge flow. The proposed isomer ratios in fresh water vapor should be verified by terahertz spectroscopy.

Our results suggest that the conventional treatment of evaporating water vapor as a single gas needs to be reconsidered in confined spaces where advection is limited. There is a prospect of relating the *ortho-para* ratio of fresh vapor to hydrogen bonding in different aqueous solutions [53], thereby obtaining new insight into the role of kosmotropic and chaotropic ions in the Hofmeister series. Furthermore, magnetic field may prove to be useful for applications where it is desirable to increase the evaporation rate of water in microscale porous media without raising the temperature [61].

Acknowledgements The work was supported by the European Commission from Contract No 766007 for the 'Magnetism and Microfluidics' Marie Curie International Training Network. JMDC acknowledges support from Science Foundation Ireland, contract 12/RC/2278_P2 AMBER. JQ acknowledges support from Science Foundation Ireland, contracts 16/RI/3403 and 17/CDA/4704. We are grateful to Luke Coburn-Moran for some of the data in Fig. 8.4 and to Anup Kumar for measurements on heavy water.

References

1. A. Abragam: The Principles of Nuclear Magnetism. Clarendon Press (1961)
2. H.B. Amor et al., Effect of magnetic treatment on surface tension and water evaporation. Int. J. Adv. Ind. Eng. **5**(3), 119–124 (2013)
3. J.S. Baker, S.J. Judd, Magnetic amelioration of scale formation. Water Res. **30**(2), 247–260 (1996)
4. H. Ben Amor et al., Experimental study and data analysis of the effects of ions in water on evaporation under static magnetic conditions. Arab. J. Sci. Eng. **47**(5), 5547–5553 (2022)
5. E. Bormashenko, Moses effect: Physics and applications. Adv. Coll. Interface. Sci. **269**, 1–6 (2019)
6. C. Buffone et al., Marangoni convection in evaporating meniscus with changing contact angle. Exp. Fluids **55**(10), 1833 (2014)
7. C. Buffone, K. Sefiane, J.R.E. Christy, Experimental investigation of self-induced thermocapillary convection for an evaporating meniscus in capillary tubes using micro–particle image velocimetry. Phys. Fluids **17**(5), 052104 (2005)
8. M.V. Carbonell, E. Martinez, J.E. Díaz, Evaporation of magnetically treated water and NaCl solutions. Int. Agrophys. **16**(3), 171–175 (2002)
9. M. Chaplin, Ortho-Water and Para-Water (2021). Available from: https://water.lsbu.ac.uk/water/ortho_para_water.html
10. E. Chibowski, A. Szcześ, Magnetic water treatment–A review of the latest approaches. Chemosphere **203**, 54–67 (2018)
11. E. Chibowski, A. Szcześ, L. Hołysz, Influence of magnetic field on evaporation rate and surface tension of water. Colloids Interfaces **2**(4), 68 (2018)
12. R. Cini, M. Torrini, Temperature dependence of the magnetic susceptibility of water. J. Chem. Phys. **49**(6), 2826–2830 (1968)
13. J.M.D. Coey: Magnetism and Magnetic Materials. Cambridge University Press (2010)
14. J.M.D. Coey, Magnetic water treatment-how might it work? Phil. Mag. **92**(31), 3857–3865 (2012)
15. J.M.D. Coey, S. Cass, Magnetic water treatment. J. Magn. Magn. Mater. **209**(1–3), 71–74 (2000)
16. V.V. Datsyuk, O.R. Pavlyniuk, Maxwell stress on a small dielectric sphere in a dielectric. Phys. Rev. A **91**(2), 023826 (2015)
17. A. Deblais et al., Viscous effects on inertial drop formation. Phys. Rev. Lett. **121**(25), 254501 (2018)
18. R. Demichelis et al., Stable prenucleation mineral clusters are liquid-like ionic polymers. Nat. Commun. **2**(1), 1–8 (2011)
19. P. Dhar, Thermofluidic transport in droplets under electromagnetic stimulus: A comprehensive review. J. Indian Inst. Sci. **99**(1), 105–119 (2019)
20. H.K. Dhavaleswarapu et al., Experimental investigation of steady buoyant-thermocapillary convection near an evaporating meniscus. Phys. Fluids **19**(8), 082103 (2007)
21. J. Dodoo, A. Stokes, Field-induced shaping of sessile paramagnetic drops. Phys. Fluids **32**, 061703 (2020)
22. J. Dodoo, A.A. Stokes, Shaping and transporting diamagnetic sessile drops. Biomicrofluidics **13**(6), 064110 (2019)
23. J.A. Dueñas et al., Magnetic influence on water evaporation rate: An empirical triadic model. J. Magn. Magn. Mater. **539**(July), 168377 (2021)
24. J.A. Dueñas et al., Effect of low intensity static magnetic field on purified water in stationary condition: Ultraviolet absorbance and contact angle experimental studies. J. Appl. Phys. **127**(13), 133907 (2020)
25. Engineering ToolBox, Evaporation from a Water Surface (2004). Available from: https://www.engineeringtoolbox.com/evaporation-water-surface-d_690.html
26. Y. Fujimura, M. Iino, The surface tension of water under high magnetic fields. J. Appl. Phys. **103**(12), 124903 (2008)

27. D. Gebauer, A. Völkel, H. Cölfen, Stable prenucleation calcium carbonate clusters. Science **322**(5909), 1819–1822 (2008)
28. Y.Z. Guo et al., Evaporation rate of water as a function of a magnetic field and field gradient. Int. J. Mol. Sci. **13**(12), 16916–16928 (2012)
29. I.M. Hauner et al., The dynamic surface tension of water. J. Phys. Chem. **8**(7), 1599–1603 (2017)
30. K. Higashitani et al., Effects of a magnetic field on the formation of $CaCO_3$ particles. J. Colloid Interface Sci. **156**(1), 90–95 (1993)
31. L. Hołysz, E. Chibowski, A. Szcześ, Influence of impurity ions and magnetic field on the properties of freshly precipitated calcium carbonate. Water Res. **37**(14), 3351–3360 (2003)
32. L. Holysz, A. Szczes, E. Chibowski, Effects of a static magnetic field on water and electrolyte solutions. J. Colloid Interface Sci. **316**(14), 996–1002 (2007)
33. D.A. Horke et al., Separating para and ortho water. Angew. Chem. Int. Ed. **53**(44), 11965–11968 (2014)
34. P. Jafari, A. Amritkar, H. Ghasemi, Temperature discontinuity at an evaporating water interface. J. Phys. Chem. C **124**(2), 1554–1559 (2020)
35. V. Jaiswal et al., Magnetohydrodynamics- and magnetosolutal-transport-mediated evaporation dynamics in paramagnetic pendant droplets under field stimulus. Phys. Rev. E **98**(1), 13109 (2018)
36. A. Kaushal et al., Soluto-thermo-hydrodynamics influenced evaporation kinetics of saline sessile droplets. Eur. J. Mech. B. Fluids **83**, 130–140 (2020)
37. J.B. Keller, M.J. Miksis, Surface tension driven flows. SIAM J. Appl. Math. **43**(2), 268–277 (1983)
38. S. Knez, C. Pohar, The magnetic field influence on the polymorph composition of $CaCO_3$ precipitated from carbonized aqueous solutions. J. Colloid Interface Sci. **281**(2), 377–388 (2005)
39. S. Kobe et al., Control over nanocrystalization in turbulent flow in the presence of magnetic fields. Mater. Sci. Eng. C **23**(6–8), 811–815 (2003)
40. F. Lafta Rashid et al., Increasing water evaporation rate by magnetic field. Int. Sci. Investig. J. **2**(3), 61–68 (2013)
41. H.E.L. Madsen, Influence of magnetic field on the precipitation of some inorganic salts. J. Cryst. Growth **152**(1–2), 94–100 (1995)
42. H.E.L. Madsen, Crystallization of calcium carbonate in magnetic field in ordinary and heavy water. J. Cryst. Growth **267**(1–2), 251–255 (2004)
43. H.E.L. Madsen, Theory of electrolyte crystallization in magnetic field. J. Cryst. Growth **305**(1), 271–277 (2007)
44. A.A. Mamrashev et al., Detection of nuclear spin isomers of water molecules by Terahertz time-domain spectroscopy. IEEE Trans. Terahertz Sci. Technol. **8**(1), 13–18 (2018)
45. Y. Nagata, K. Usui, M. Bonn, Molecular mechanism of water evaporation. Phys. Rev. Lett. **115**(23), 1–5 (2015)
46. J. Nakagawa et al., Magnetic field enhancement of water vaporization. J. Appl. Phys. **86**(5), 2923–2925 (1999)
47. S.M. Pershin, A.F. Bunkin, Temperature evolution of the relative concentration of the H_2O ortho/para spin isomers in water studied by four-photon laser spectroscopy. Laser Phys. **19**(7), 1410–1414 (2009)
48. J.S. Philo, W.M. Fairbank, Temperature dependence of the diamagnetism of water. J. Chem. Phys. **72**(8), 4429–4433 (1980)
49. E.M. Pouget et al., The initial stages of template-controlled $CaCO_3$ formation revealed by cryo-TEM. Science **323**(5920), 1455–1458 (2009)
50. S. Poulose et al., Magnetic field enhances evaporation of water in confined spaces. IEEE Magn. Lett. **2023**(11), 1–5 (2020)
51. S. Poulose et al., Deformation and necking of liquid droplets in a magnetic field. Phys. Fluids **34**(11), 112116 (2022)

52. S. Poulose et al., Evaporation of water and urea solution in a magnetic field; the role of nuclear isomers. J. Colloid Interface Sci. **629**, 814–824 (2023)
53. B. Rana, D.J. Fairhurst, K.C. Jena, Investigation of water evaporation process at air/water interface using Hofmeister ions. J. Am. Chem. Soc. **144**(39), 17832–17840 (2022)
54. R.E. Rosensweig: Ferrohydrodynamics. Dover Books (2013)
55. A. Seyfi, R. Afzalzadeh, A. Hajnorouzi, Increase in water evaporation rate with increase in static magnetic field perpendicular to water-air interface. Chem. Eng. Process. **120**(January), 195–200 (2017)
56. M.D. Simon, A.K. Geim, Diamagnetic levitation: Flying frogs and floating magnets (invited). J. Appl. Phys. **87**(9), 6200–6204 (2000)
57. S.J. Suresh, V.M. Naik, Hydrogen bond thermodynamic properties of water from dielectric constant data. J. Chem. Phys. **113**(21), 9727–9732 (2000)
58. A. Szcześ et al., Effects of static magnetic field on water at kinetic condition. Chem. Eng. Process. **50**(1), 124–127 (2011)
59. G. Verma, K.P. Singh, Time-resolved interference unveils nanoscale surface dynamics in evaporating sessile droplet. Appl. Phys. Lett. **104**, 1–4 (2014)
60. Y. Wang, H. Wei, Z. Li, Effect of magnetic field on the physical properties of water. Results Phys. **8**, 262–267 (2018)
61. K. Yang, G. Ye, G. de Schutter: Will ortho-enriched water increase the durability of concrete. In: Proceeding of 4th International RILEM Conference on Microstructure-Related Durability of Cementitious Composites, pp. 713–720 (2020)
62. Q.W. Yang, H. Wei, Z. Li, Enhancing water evaporation by combining dynamic and static treatment of magnetic field. Desalin. Water Treat. **216**, 299–305 (2021)
63. C. Zhang et al., Hydrophobic treatment on polymethylmethacrylate surface by nanosecond-pulse DBDs in CF4 at atmospheric pressure. Appl. Surf. Sci. **311**, 468–477 (2014)

Chapter 9
Influence of Large Magnetic Field Gradients at the Electrochemical Interface

**Jinu Kurian, Peter Dunne, Vincent Vivier, Gwenaël Atcheson,
Ruslan Salikhov, Ciaran Fowley, Munuswamy Venkatesan, Olav Hellwig,
Michael Coey, and Bernard Doudin**

9.1 Introduction

There is renewed interest in the importance of magnetic fields and magnetic materials on (electro)chemical reactions [1, 2], driven by possible energy applications, ranging from charge storage, to solar cells and hydrogen production [3]. The situation is complex, with multiple processes simultaneously at play, ranging from microhydrodynamics and magneto-chemistry to spin-dependent surface chemistry [4, 5]. As a result, experiments are mostly performed by trial-and-error. Nearly all experiments make use of large permanent magnets or are conducted within the bore of large-scale solenoid flux sources. The region between electrode and electrolyte can be divided into two; a few nm thin region of electrolyte at the electrode known as double layer,

J. Kurian (✉) · P. Dunne · B. Doudin
CNRS, IPCMS UMR 7504, Université de Strasbourg, 23 Rue du Loess, 67034 Strasbourg, France
e-mail: jinukurian9048@gmail.com

P. Dunne
e-mail: peter.dunne@ipcms.unistra.fr

B. Doudin
e-mail: bernard.doudin@ipcms.unistra.fr

V. Vivier
CNRS, Laboratoire de Réactivité de Surface, UMR 7197, Sorbonne Université, 4 Place Jussieu, 75005 Paris, France
e-mail: vincent.vivier@sorbonne-universite.fr

G. Atcheson · M. Venkatesan · M. Coey
AMBER and School of Physics, Trinity College, Dublin, Ireland
e-mail: atchesog@tcd.ie

M. Venkatesan
e-mail: venkatem@tcd.ie

© The Author(s) 2024   111
B. Doudin et al. (eds.), *Magnetic Microhydrodynamics*, Topics in Applied Physics 120,
https://doi.org/10.1007/978-3-031-58376-6_9

or Stern layer, which then extends as the microns to mm thick diffusion layer. Most of the magnetic field effects influence the diffusion layer, which controls the mass transport in the electrochemical cell [2, 3] and only indirectly impact the diffuse part of the Stern layer [6]. However, key electrochemical reactions occur at short distance from the electrode surface (Fig. 9.1), and specifically within the first few nanometers of solvents and reactants, in the double layer that develops at the liquid (electrolyte)-solid (electrode) interface [6, 7]. Ideally, one would therefore like to probe magnetic effects by applying intense magnetic fields confined to only a short distance from the electrode surface.

A magnetic field can influence an electrochemical reaction mainly by means of the Lorentz force [8] and the Kelvin force [9]. The Lorentz force, which can be expressed as

$$F_L = j X B \qquad (9.1)$$

will induce convection when the current density $\vec{j}$ and magnetic field $\vec{B}$ are non-collinear. In a typical electrochemical system with $B = 1$ T and $j = 10^3$ A/m^2, the Lorentz force will be of the order of 10^3 N/m^3, comparable to the buoyancy-driven convective force. This magnetohydrodynamic (MHD) force modifies the convection, resulting in compression of the diffusion layer, which is similar to the effect of mechanical stirring. However, due to its dependence on the local current density, MHD effect can generate edge-effect induced vorticity [10], flow patterns in electrodeposits [11], and can even influence bubble formation due to micro-MHD flows [12].

On the other hand, the magnetic field gradient force can be expressed as

$$F_{\nabla B} = \frac{1}{2\mu_0} c \chi_m (B \cdot \nabla) B, \qquad (9.2)$$

M. Coey
e-mail: jcoey@tcd.ie

R. Salikhov · C. Fowley · O. Hellwig
Institute of Ion Beam Physics and Materials Research, Helmholtz-Zentrum Dresden-Rossendorf, Bautzner Landstraße 400, 01328 Dresden, Germany
e-mail: r.salikhov@hzdr.de

C. Fowley
e-mail: c.fowley@hzdr.de

O. Hellwig
e-mail: o.hellwig@hzdr.de

O. Hellwig
Institute of Physics, Chemnitz University of Technology, Reichenhainer Strasse 70, 09107 Chemnitz, Germany

$\sim 1 - 10$ nm

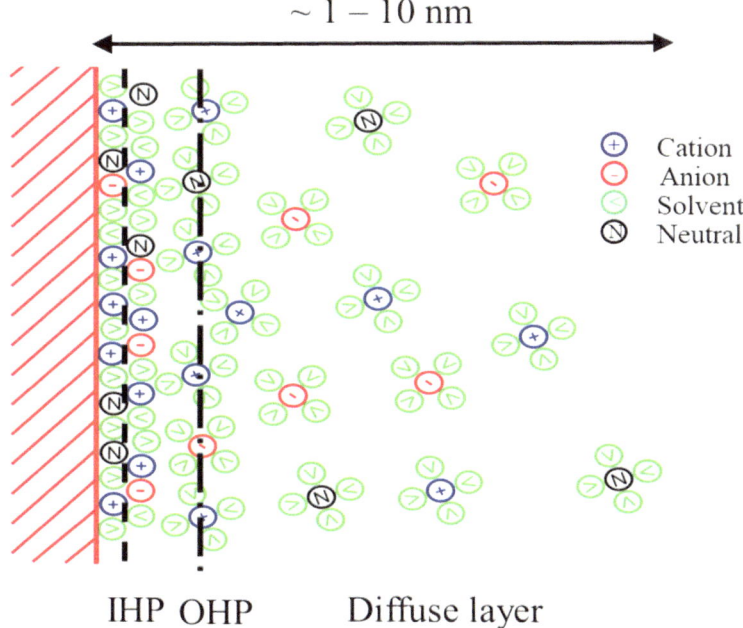

$\oplus$	Cation
$\ominus$	Anion
$\oslash$	Solvent
$\oslash$	Neutral

IHP OHP   Diffuse layer

Fig. 9.1 Schematic of the double-layer structure close to an electrode surface; the Stern layer consists of compactly arranged solvated ions and neutral molecules, which forms the inner and outer Helmholtz planes (IHP, OHP) and a diffuse layer due to the thermal agitation in the solution. Reprinted with permission from [6]. Copyright 2019 American Chemical Society

where $\mu_0 = 4\pi \, 10^{-7} H/m$ and c is the concentration of paramagnetic species having a small enough molar magnetic susceptibility χ_m (as always encountered in electrochemical solutions). This field gradient force, proportional to the product of the scale-independent magnetic field and its gradient, will be active whenever there is a concentration gradient that is not parallel to the non-homogenous field [8, 13, 14]. In an electrolyte containing paramagnetic ions with $c = 1$ M, $\chi_m = 10^{-8}$ m^3/mol, $\mathbf{B} = 1$ T and $\nabla B = 100$ T/m, the magnetic field gradient force is of the order of 10^3 N/m^3, comparable to natural convective forces. It has been shown that the magnetic field gradient force can levitate diamagnetic materials in paramagnetic media [15], influence the efficiency of the hydrogen reduction [16], electrodeposit direct and inverse patterns [17–19], and can even stabilize liquid based frictionless microfluidic channels [20, 21].

Reduction of the size of magnetic sources to the nanoscale opens new possibilities, with enhanced field gradient forces and broadband control that is orders of magnitudes larger than currently reported in the literature. We therefore focus on increasing one of the two magnetic forces at play, namely the magnetic field gradient force originating from the spatial derivative of the magnetic field, which increases as we decrease the size of the system. This gain in force magnitude is obtained by generating a field and field gradient acting on a very small volume about 10 nm region at the interface between a metallic film and a solution, exactly where it will have the greatest impact on electrochemical reactions. Our approach relies on the use of magnetic thin films stacks, made of alternating Co and Pt layers, optimized to keep their magnetization normal to the plane of the films. We therefore take advantage of technological advances developed for perpendicular magnetic recording applications. We investigate how these planar electrodes compare to pure Pt films for a benchmark single e^- exchange reaction. Our strategy is therefore to simplify to the maximum the type and location of the magnetic force at play on a benchmark electrochemical reaction process.

9.2   Experimental Methods

Magnetic thin film stacks of Ta(5.0)/Pt(2.0)/ [Co(0.8)/Pt(0.8)]$_N$/Pt(3.0) [numbers in nm] with N = 10, 20 and 50 multilayers were DC magnetron sputtered onto thermally oxidized Si substrates. The Ta/Pt in the stack served as a buffer layer for the growth of Co/Pt repeats while the outer Pt functioned as a capping layer to prevent oxidation and maintain a stable surface. Films consisting just of Ta(2.0)/Pt(20.0) are also sputter deposited as a non-magnetic film benchmark. All multi-repeat growth thicknesses were confirmed using X-ray reflectivity measurements (not shown here). All Co/ Pt based samples are magnetically characterized using superconducting quantum interference device (SQUID) magnetometry and magnetic force microscopy (MFM).

A three-electrode system was used to characterize the electrochemical ferricyanide/ferrocyanide redox reaction. This is a well-documented single e^- redox reaction with an appropriate energy and kinetic window for experiments that is widely used as a standard for testing electrochemical systems [22, 23]. Pt or Co/ Pt films were used as the working electrode (WE), Pt mesh as the counter electrode (CE) and an Ag/AgCl electrode in 3 M KCl as the reference electrode (RE). The WE area was typically 1–10 mm^2, defined using electro-inactive epoxy masking the sides of the thin film sample. This protected the thin film electrode against oxidation and ensured the absence of magnetic fringe field effects (where we do not control the magnitude and direction of the stray field) that could perturb the interpretation of the results if the sides of the magnetic electrode are exposed to the solution. The electrolyte, made of 0.2 M potassium ferricyanide (K$_3$[Fe(CN)$_6$], > 98% purity) as electroactive species and 1 M potassium chloride (KCl, 99% purity) as supporting electrolyte, was freshly prepared prior to measurements.

Impedance spectra were measured under potentiostatic conditions, applying a 10 mV perturbation signal having frequency in the range from 1 Hz to 100 kHz. Linearity of the data obtained was confirmed using Kramers–Kronig transformations. A Pt wire coupled in parallel with the reference electrode through a 0.1 μF capacitor was used to avoid high frequency artifacts [24]. Impedance data were fitted with a Randles circuit composed of a charge transfer resistance (R_{ct}) in series with a semi-infinite Warburg diffusion resistance (Z_w), both in parallel with the double layer capacitance expressed as a constant phase element (CPE), and all in series with the solution resistance which includes uncompensated as well as cable resistances. The effective double layer capacitance was estimated from the CPE element using the relation [25, 26],

$$C_{eff} = P^{1/\alpha}\left[\frac{1}{R_s} + \frac{1}{R_{ct}}\right]^{(\alpha-1)/\alpha}, \qquad (9.3)$$

where P is related to the CPE impedance as $Z(f) = 1/[P(j2\pi f)^{\alpha}]$ where f and α are frequency and distribution factors respectively. The solution resistance was obtained before each measurement by a positive feedback method and agrees with the resistance value obtained from impedance spectroscopy. Cyclic voltammetry measurements were obtained after correcting for the ohmic drop, $i R_s$. Chronoamperograms were measured at potential of 0.25 V after waiting 10 min to reach the steady state.

9.3 Magnetic Source Design

Co/Pt films are best known for their perpendicular magnetic anisotropy; their magnetic properties and magnetization switching are well documented in the literature [27–29]. In the present study, magnetic force microscopy (MFM) is used to image the magnetic domains of the thin films in their remanent state. Figure 9.2 shows the multi-state maze-like domain patterns of Co/Pt films with different number of Co repeats. To estimate the domain width, the contrast difference across stripes at different positions is fitted using a Gaussian function and the FWHM of the fit is taken as the average domain size. A minimum domain width D of 130 ± 25 nm is found when N = 20 (Table 9.1). The change of domain width with number of Co/Pt repeats can be well understood within the model of periodic stripe domains with uniaxial magnetic anisotropy developed by Kittel [30, 31] and further extended by Kooy and Enz [32]. According to this model, the total magnetic energy at remanence can be expressed in terms of magnetostatic energy and a domain wall energy, the balance of which determines the domain width. Similar observation of variation of domain width with number of repeats has been reported by Diao et al. [33] and Hellwig et al. [28].

Table 9.1 summarizes the relevant parameters characterizing our Co/Pt thin films. The anisotropy field, H_k, is defined as the field required to saturate the sample along its hard axis. Considering the perpendicular magnetic anisotropy of the Co/Pt films, H_k can be approximated as the in-plane field required to saturate the magnetization parallel to the film surface. The effective anisotropy can be found using the relation,

$$K_{eff} = \frac{\mathbf{M_s} \cdot \mathbf{H_k}}{2}. \qquad (9.4)$$

In terms of volume anisotropy K_v and surface anisotropy K_s,

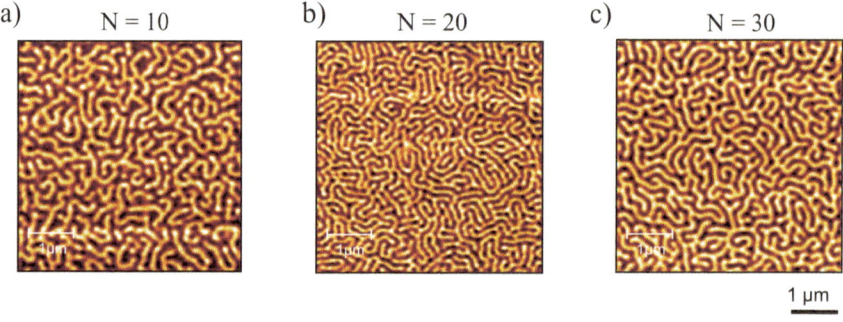

Fig. 9.2 MFM images (5.0 μm × 5.0 μm) of [Co(0.8)/Pt(0.8)]$_N$ multilayer samples at tip lift height 50 nm with **a** N = 10, **b** N = 20 and **c** N = 50 Co/Pt repeats in the remanent state after out-of-plane saturation

$$K_{eff} = K_v + \frac{2K_s}{t} \tag{9.5}$$

where t is the magnetic film thickness. The multiplier 2 is because there are two interfaces on either side of the thin film. Using Kittel's magnetic domain model, the Bloch wall width can be estimated as

$$\lambda_w = \pi \sqrt{A/K_u} \tag{9.6}$$

where A is the exchange stiffness, taken to be about 10 pJ/m [34] and K_u is related to surface anisotropy by $K_u = K_s/t$. Thus, the domain wall thickness is found to be of the order of 5–10% of the domain width, depending on the number N of Co/Pt repeats. These results are in line with more detailed studies on similar multilayers in the literature [35, 36].

Estimates of the magnitude of the stray magnetic field and the related field gradient as a function of distance from the film surface are key for understanding the field effects at the electrochemical interface. To calculate the field magnitude, thin films are considered as 2D sheets magnetized normal to surface, defined along the *z-axis*. Furthermore, individual domains are assumed to be infinitely long and rectangular

Table 9.1 Summary of relevant parameters of Co/Pt films. The saturation magnetization of the Co layer M_s and the anisotropy H_k of the Co/Pt stack are obtained using SQUID magnetometry

N	Total t_{Co} (nm)	$M_s \pm 0.03$ (MA/m)	$H_k \pm 0.03$ (T)	K_{eff} (kJ/m^3)	D (nm)	λ_w (nm)
10	8	1.53	0.71	535 ± 14	250 ± 40	12.5
20	16	1.52	0.79	600 ± 17	130 ± 25	12.1
50	40	1.54	0.97	746 ± 23	200 ± 25	11.5

The domain size D is obtained from MFM images analysis and the domain wall width λ_w is calculated using Eq, (9.6)

and the field components can be analytically derived using the Amperian current model as [37],

$$\mathbf{B} = B_x \hat{x} + B_z \hat{z} \tag{9.7}$$

$$B_x = \frac{\mu_0 M_r}{4\pi} \left[\ln\left(\frac{(x+a)^2 + (z-b)^2}{(x+a)^2 + (z+b)^2} \right) - \ln\left(\frac{(x-a)^2 + (z-b)^2}{(x-a)^2 + (z+b)^2} \right) \right] \tag{9.8}$$

$$B_z = \frac{\mu_0 M_r}{2\pi} \left[\arctan\left(\frac{2b(x+a)}{z^2 - b^2 + (x+a)^2} \right) - \arctan\left(\frac{2b(x-a)}{z^2 - b^2 + (x-a)^2} \right) \right] \tag{9.9}$$

where $2a$ is the width of domain and $2b$ is the thickness of the magnetic layer. A python code for such calculation is available online [38]. In a magnetic film, the sources of the magnetic stray field are the edges of the magnetic domains. Hence, a monodomain film with minimal edges would generate a smaller field gradient (near zero) compared to the large near-surface field gradient generated by the multi-domain magnetic films. In such a state, the field gradient is maximum near the domain wall where the magnetization direction changes rapidly. Figure 9.3a shows a contour plot of the field gradient generated by a $[Co/Pt]_{20}$ film with a large resulting field gradient of the order of $10^6 - 10^7$ T/m generated by the multi-domain magnetic structure near its surface. Figure 9.3b shows the average field gradient values for three different repeats. It is found that $N = 20$ multilayer generates a field gradient of magnitude comparable to that of the $N = 50$ multilayer and it is expected that the gradient field effects will be more localized near the surface of N=20 multilayer as the values drop faster compared to that of the $N = 50$ sample. $N = 20$ multilayers are therefore chosen as the magnetic working electrode to investigate the field induced effects presented in the electrochemistry experiments. It is worthwhile to note that these field gradient values can have local maxima much larger than the plotted average values of the gradient magnitude along the x-axis as a function of the distance from sample surface shown in Figure 9.3b. Furthermore, this field calculation is based on the assumptions such as (i) domain wall regions with tilted magnetization are considered to be magnetically dead, (ii) equal-width domains; and (iii) a perfect interface and a flat surface.

9.4 The Electrochemical System

The one e^- transfer outer sphere ferricyanide/ferrocyanide redox couple is chosen as the model electrochemical system. It is commonly used as a standard redox probe due to its electrochemical reversibility [39] and stability [40]. The reaction is:

$$\left[Fe(CN)_6 \right]^{4-} \leftrightarrow \left[Fe(CN)_6 \right]^{3-} + e^-, \tag{9.10}$$

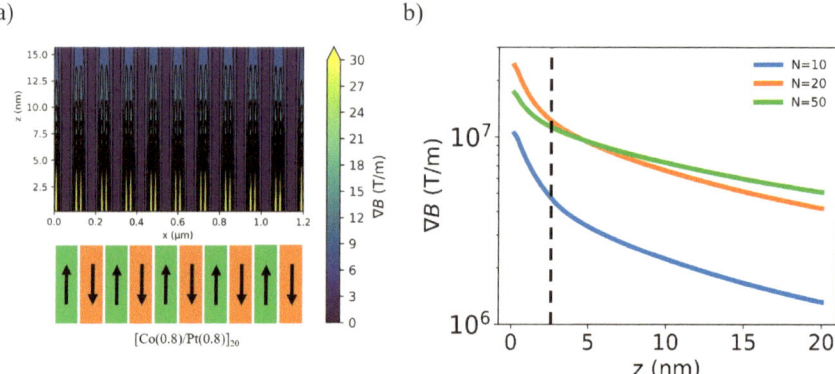

Fig. 9.3 **a** Near-surface magnetic field gradient contour plot calculated for the [Co/Pt]$_{20}$ multilayer. The schematics represents the cross-sectional view of Co/Pt repeats with gap between the alternative domains corresponding to the dead region. Relevant parameters: magnet thickness = 16.0 nm, domain size = 0.13 μm, magnetization = 1.54 MA/m, magnetically dead region = 9%. **b** line plot of the average field gradient magnitude as a function of distance from the film surface for the three stacks described in Table 9.1

where the ferrocyanide anion, $[Fe(CN)_6]^{4-}$, is reversibly oxidized to form the ferricyanide anion, $[Fe(CN)_6]^{3-}$. Ferrocyanide is diamagnetic with a low spin ferrous Fe^{2+} while ferricyanide is paramagnetic with iron in the high spin ferric Fe^{3+} state. In this redox reaction, both species are soluble in water and remain as solvated ions, minimizing the changes to the electrode surface during electrochemical measurements, unlike electrodeposition or corrosion. The reaction's redox potential ensures that gas generation due to water splitting (hydrogen or oxygen evolution) is negligible during electrochemical measurements, thus minimizing micro-convective effects due to bubble formation [41].

Prior to the effects induced by magnetic field, it is important to characterize the properties of the reaction in detail. Cyclic voltammetry was used to estimate the kinetics as well as the diffusion parameters. It is known that the kinetics of the $[Fe(CN)_6]^{3-}/[Fe(CN)_6]^{4-}$ electrode process is highly dependent on the cleanliness of the electrode [42]. In the present study, the redox peaks in the cyclic voltammogram are found to be absent when either Pt or Co/Pt films are used as WE without any prior cleaning. Various cleaning procedures have been reported in the literature to improve the electrode activity [43]. We followed the procedure described in [42], which starts with sonication of the thin film in acetone for 2 min followed by a rinse using ethanol and isopropanol. For the electrochemical cleaning, the potential was swept between −1.0 V and +1.3 V at a scan rate of 100 mV/s in 0.1 M KCl solution until a steady state response was observed.

After WE pretreatment, the ferricyanide/ferrocyanide redox reaction is characterized by sweeping the potential and measuring the current response under a uniform surface normal field of 400 mT so that the Pt electrode properties can be compared to those of the Co/Pt electrodes in their saturated mono-domain state (Figure 9.4a). The

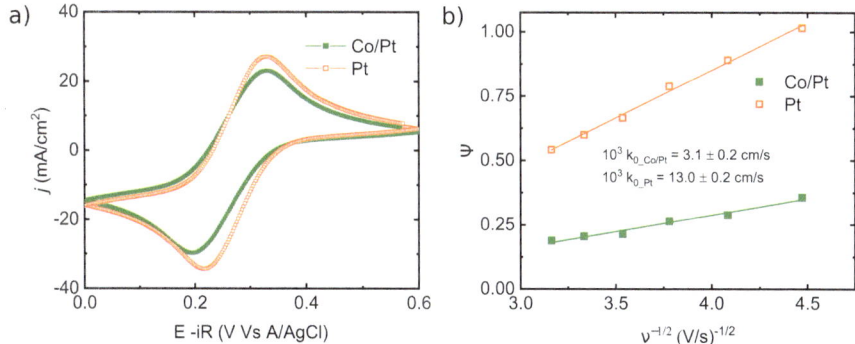

Fig. 9.4 **a** Cyclic voltammogram at a scan rate of 50 mV/s and **b** dimensionless parameter Ψ of Pt and Co/Pt electrode in solution containing 0.2 M $K_3[Fe(CN)_6]$ and 1 M KCl

solution resistance, R_s, is found to be 21 Ω using a positive feedback technique. The open circuit potential (OCP) of the system is measured as 0.41 V. When the potential is swept from 0.6 to 0.0 V, reduction of ferricyanide occurs while the oxidation of reduced species occurs in the reverse sweep. According to the Randles–Sevcik equation for a reversible system of the type $O + ne^- \rightarrow R$, the peak current at room temperature follows the relation, which holds for low scan rate only,

$$i_{p,c} = -2.69 - 2.6910^5 n^{\frac{3}{2}} A D_0^{\frac{1}{2}} [O]_\infty v^{1/2}, \tag{9.11}$$

where v is the scan rate in V/s, A is electrode area in cm^2, D is analyte diffusion coefficient in cm^2/s, $[O]_\infty$ is the bulk analyte concentration in mol/cm^3, and n is the number of electrons transferred in the redox reaction. In order to estimate the diffusion coefficient, i_p is plotted against $v^{1/2}$ and fitted assuming a linear relation (Eq. 9.11). Irrespective of the electrode, the diffusion coefficient of the ferricyanide reduction reaction is found to be ~ 6×10^{-6} cm^2/s and that of ferrocyanide oxidation is ~ 5×10^{-6} cm^2/s. As the diffusion coefficient indicates the quantity of diffused particles per unit time, the difference is ascribed to the lower density of the ferricyanide compared to that of the ferrocyanide [44]. The obtained values are comparable to the literature values $5–7.2 \times 10^{-6}$ cm^2/s for ferricyanide and $4.5–6.4 \times 10^{-6}$ cm^2/s for ferrocyanide [45, 46].

For high scan rates, the electrochemical system behaves as a quasi-reversible one, and the separation between redox peaks in CV curves is measured to get insight into the kinetic rate constant. At a scan rate of 50 mV/s, the peak separation ΔE_p for a Pt electrode is found to be 77 mV while that of a Co/Pt electrode is 98 mV. A higher ΔE_p value compared to the 57 mV separation for a reversible reaction can be related to the slow kinetics [47] or electrode surface contamination [48]. The heterogeneous standard kinetic rate constant k_0 can be extracted from the ΔE_p dependence on the scan rate v using the empirical relation [49],

$$\psi = (-0.6288 + 0.0021X)/(1 - 0.017X), \qquad (9.12)$$

where $X = \Delta E_p \times n$, the rate constant k^0 can be obtained using the Nicholson approach [47] where ψ is given by

$$\psi = k^0/[\pi \, DnFv/RT]^{1/2}, \qquad (9.13)$$

To estimate the standard rate constant, ψ is calculated using Eq. (9.12) and is plotted against $v^{-1/2}$ (Fig. 9.4b). Fitting the plot with a linear function and using Eq. (9.13), k0 of Co/Pt and Pt electrodes are estimated as 3.1×10^{-3} and 13.0×10^{-3} cm/s, respectively. The k_o-value obtained for the Pt film is within the wide range 0.01–0.4 cm/s reported for Pt electrode based ferricyanide/ferrocyanide systems [48, 50]. The lower rate constant for Co/Pt could be attributed to a non-perfect coverage of Pt as top layer of the magnetic stack. The dependence of k_0 on the magnetic field will be discussed in the next section.

9.5 Magnetic Field Effects on the Electrochemical Reaction

The impact of a magnetic field on Pt and Co/Pt film electrodes is studied using cyclic voltammetry in two ways (Fig. 9.5a, b) the external field is applied out of the plane of the film (OP) or parallel to the current density, and (c) (d) the field is applied in the plane of the film (IP) or perpendicular to the current density. Figure 9.5a confirms that there are no field-induced effects in the OP field geometry on a Pt benchmark electrode. Thus, it is safe to assume that micro-MHD and edge-effect, induced field driven convective effects are minimized in the system in this magnetic configuration. In the case of the Co/Pt electrode (Fig. 9.5b), a small change in reduction current can be observed with the perpendicular field which might be due to localized field gradient effects (see the discussion in the next section). However, no appreciable shift in peak separation is observed. It confirms that the observed difference in k_o between Pt and Co/Pt electrodes is likely to be a chemical surface effect. We also checked that upon covering the Co/Pt with a thicker 10 nm Pt cap layer, the k_o increased to 10×10^{-3} cm/s, becoming comparable to that of a pure Pt film. However, the thick Pt overlayer will severely diminish the nanoscale field gradient effects on the electrochemical interface (see Fig. 9.3b). Hence, the field effect studies were performed with 3 nm capped Co/Pt films, keeping in mind the difference in kinetic activity, but still pertinent to check if kinetics can change when modifying the magnetic state of the electrode.

Lorentz force induced effects can be probed when a uniform magnetic field is applied parallel to the electrode surface. Figure 9.5c, d illustrate the MHD effect on both Pt and Co/Pt with an IP magnetic field configuration. During reduction of ferricyanide, Lorentz force effects are negligible in a potential range 0.6–0.25 V/Ag/AgCl which corresponds to the kinetics limited region whereas the diffusion limited region (0.275–0.6 V/Ag/AgCl) is considerably influenced by the in-plane magnetic field. We can also observe from the cyclic voltammograms that the Lorentz force effects on

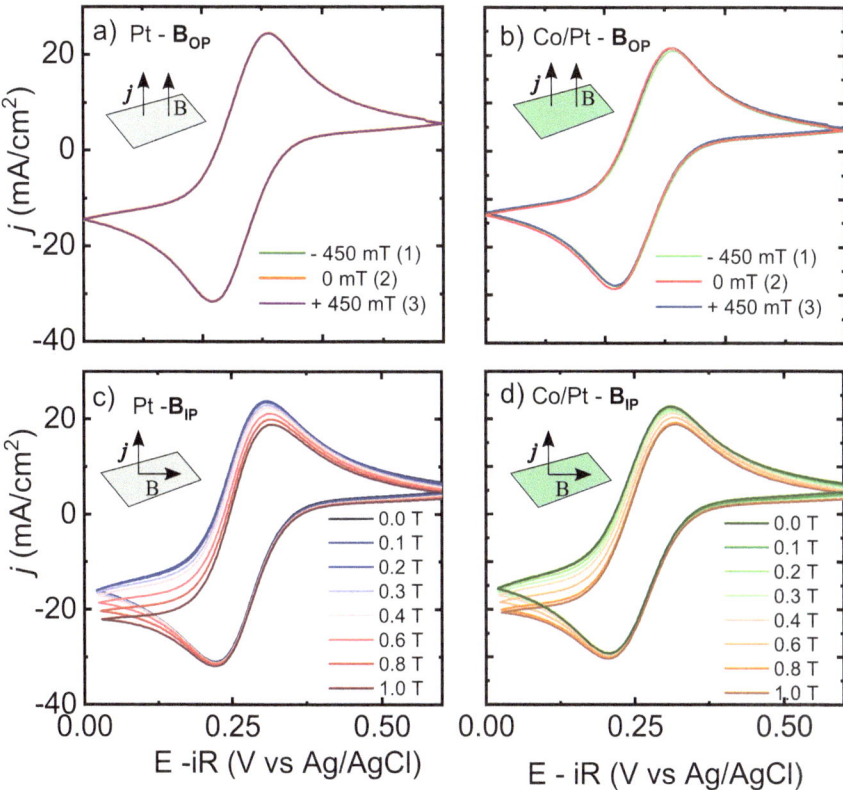

Fig. 9.5 Cyclic voltammograms of **a** Pt and **b** Co/Pt with 2.8 nm cap layer under different out-of-plane magnetic field. CV curves of **c** Pt and **d** Co/Pt film under different in-plane magnetic fields. Scan rate of all CV is 50 mV/s

the two reaction peaks are asymmetrical, which can be explained by the difference in absolute current density values. As the electrolyte consists of ferricyanide solution, the current density is higher in the reduction region (left) compared to the oxidation region (right) in Fig. 9.5c, d, resulting in a difference in the Lorentz force magnitude. Similar in-plane field behavior is observed for both Pt and Co/Pt film electrodes and therefore indicates that the magnetic domain configuration of the Co/Pt samples does not play a significant role in this field configuration.

Cyclic voltammograms give qualitative information on the mass transport response to the magnetic field. In order to better understand the Lorentz force effects on the diffusion layer, electrochemical impedance spectroscopy (EIS) is used. Figure 9.6a shows the evolution of the impedance spectra when varying the IP magnetic field magnitude at a bias voltage of 0.15 V/Ag/AgCl, where active reduction of ferricyanide is mass transport limited. The Nyquist plot consists of a semi-circular curve in the high frequency region (left), which is related to the interfacial properties and a slanted straight line in the low frequency region (right side)

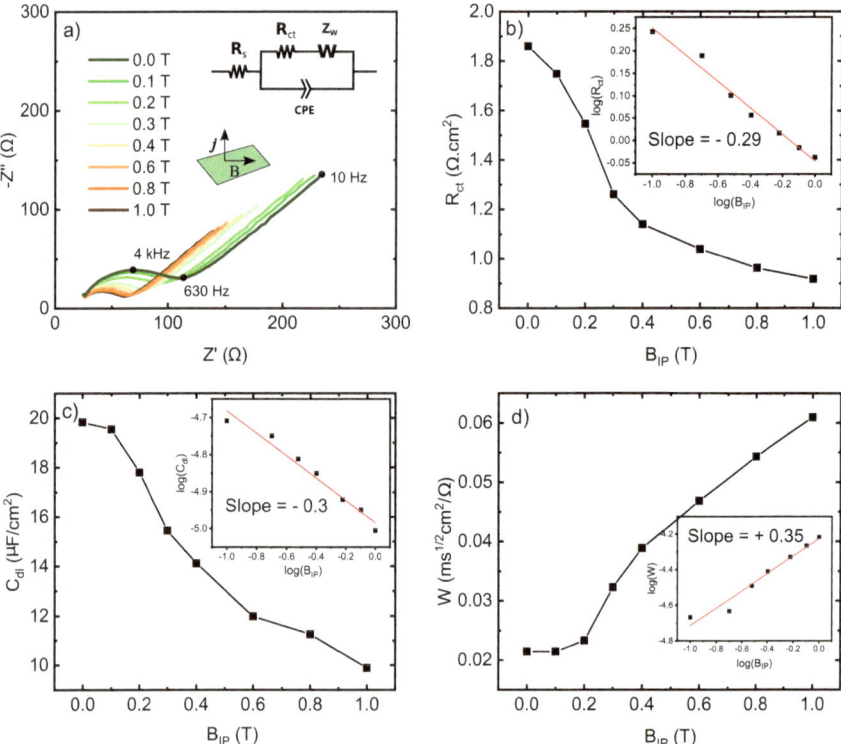

Fig. 9.6 **a** Evolution of impedance spectra of Co/Pt electrode at a bias voltage 0.15 V/Ag/AgCl under different magnetic field applied along the plane of film. Randles circuit fit parameters **b** charge transfer resistance **c** effective double layer capacitance and **d** Warburg element as a function of in-plane field

related to the diffusion layer. The Warburg element, related to the diffusion layer, is found to be very sensitive to the applied field with a power law dependence of $B^{1/3}$ (Fig. 9.6d), as predicted by Aogaki [51] for magneto-convection induced by the Lorentz force. The change of R_{ct} and C_{dl} with the IP field (Fig. 9.6b, c) suggests that both interfacial and diffusion regions are sensitive to the bulk convection. We further confirmed the convective nature of the Lorentz force by studying the reaction under forced convection induced by mechanical stirring and found a similar behavior that we do not show here.

In order to study how a large field gradient at the electrochemical interface can impact the reaction, impedance spectra of the bath using the magnetic Co/Pt WE are recorded when imposing several different OP field conditions (Fig. 9.7a), under 0.25 V/Ag/AgCl potentiostatic conditions. The data is fitted using the Randles circuit like the one for IP field measurements. The bath resistance (R_s) 19.7 ± 0.2 Ω is found to be almost independent of the applied magnetic field.

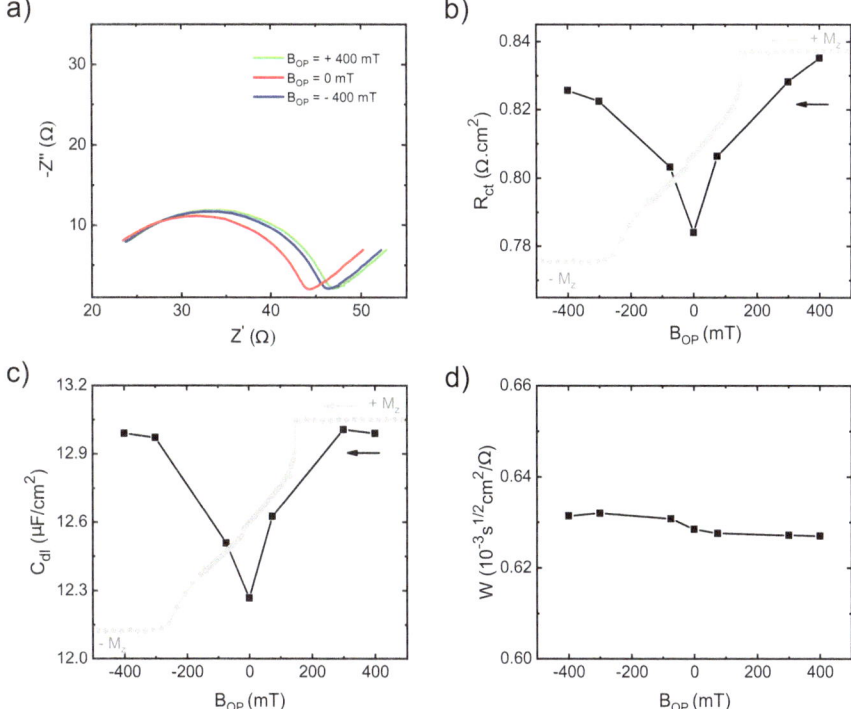

Fig. 9.7 a Evolution of the impedance spectra of Co/Pt electrode at a bias voltage 0.25 V/Ag/AgCl under different magnetic fields applied normal to the film surface. Randles circuit fit parameters **b** charge transfer resistance, **c** double layer capacitance and **d** Warburg diffusion impedance as a function of magnetic field. The magnetization evolution of $[Co(0.8)/Pt(0.8)]_{20}$ with OP field is shown in the background of figure **b** and **c** (grey)

However, a significant difference occurs in the high frequency region, where the charge transfer resistance (R_{ct}) and the double layer capacitance (C_{dl}) both show a clear dependence on the applied magnetic field (Fig. 9.7b, c). Both quantities are increased by 5–6% when the applied field changes from 0 to $\pm$ 400 mT. These values correspond to the magnetic film in the in multi-domain state (large field gradient) at 0 mT and mono-domain state (minimum field gradient) when in a $\pm$ 400 mT applied field. As the applied DC bias voltage is close to the half potential $E_{1/2}$, the effective double layer capacitance can be treated as a Gouy-Chapman capacitor corresponding to the diffusive double layer region. A change in C_{dl} with applied field would then imply a change in the concentration of ions in the diffusive double layer region. Thus, a field-gradient driven micro- near-electrode convection is assumed to be the origin of the observed changes of impedance spectroscopy.

To check if the field dependence of the EIS with a magnetic Co/Pt WE was intrinsic to the magnetic properties of the WE, measurements under identical conditions were performed using a non-magnetic Pt film as WE. Figure 9.8 shows the corresponding

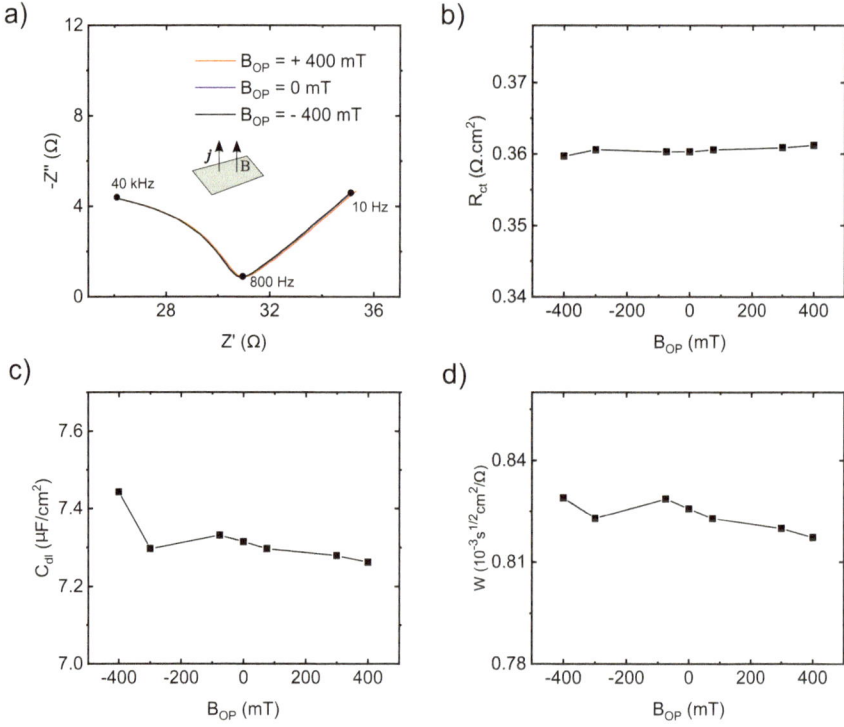

Fig. 9.8 **a** Evolution of the impedance spectra of Pt electrode at a bias voltage 0.275 V/Ag/AgCl under different magnetic field applied parallel to the current density. Randles circuit fitted parameters **b** charge transfer resistance, **c** double layer capacitance and **d** Warburg diffusion impedance as a function of magnetic field

Nyquist diagrams as well as the evolution of the Randles circuit parameters under different magnetic fields. It reveals that neither interfacial nor bulk properties of the reaction are sensitive to the applied magnetic field. Hence, the perpendicular field has almost no effect for the nonmagnetic WE. Irrespective of the field value, R_{ct} and C_{dl} remain unchanged within about 1–2%, which shows the reproducibility of the measurements. It confirms the CV data of Fig. 9.5a, with a picture of reproducible and stable reaction along with the minimum external-field induced effects in this particular configuration.

Insight into the possible magnetic effect on the charge transfer coefficient of the reaction can be gained by testing how the product of current i and the charge transfer resistance R_{ct} evolves with the change of magnetic force field amplitude [52]. The steady state current response to the OP field is therefore measured for Pt and Co/Pt films under the same potentiostatic condition (0.25 V/Ag/AgCl) as the one used for EIS measurements (Fig. 9.9). The applied OP field is swept between ± 400 mT in steps of 10 mT. Once the applied field is stabilized (in < 5 s), the current is measured after a 10 s waiting time, implying a delay time of 15 s in total between two

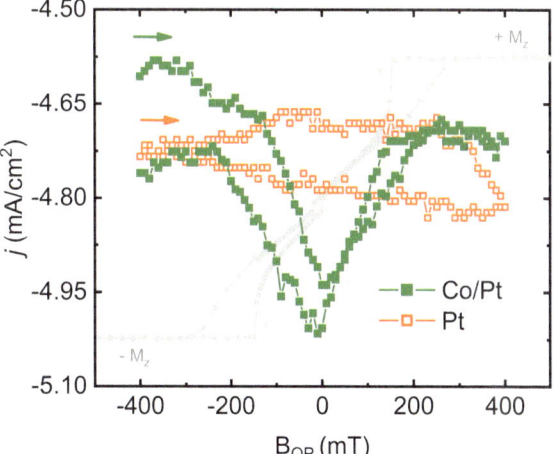

Fig. 9.9 Steady state current as a function of OP applied magnetic field, increasing when the magnetic force field generated by the Co/Pt multilayer electrode increases, and nearly invariant for a non-magnetic benchmark Pt electrode

consecutive current measurements. While the steady-state current of a Pt electrode is almost invariant with applied field, the Co/Pt electrode shows a clear change in current density with applied OP field, having a maximum magnitude at 0 mT where the film is in multi-domain state, and a minimum when the field is > |200| mT where the film magnetization becomes saturated (Fig. 9.9). Upon changing the field from ±400 to 0 mT, the current density of the Co/Pt WE increases by around 7%, which implies that the product $i*R_{ct}$ is roughly invariant under applied OP field conditions. This support the claim that the charge transfer process, or the reaction kinetics, under these specific potentiostatic conditions are not influenced by the large magnetic force at the interface.

Returning to the OP field dependence on the reaction, the micro-MHD and edge effect driven convective effects can be assumed to be negligible in the system as no field-induced effects are observed for the Pt-film WE. Hence, the external field cannot be considered as a primary source for these changes. We have therefore strong experimental evidence that the localized stray field gradient generated by the Co/Pt stack in the multi-domain state at small applied external field is responsible for the observed changes in the Randles cell components and the steady state current. This force field is expected to be acting near the immediate vicinity of the electrode, within the few tens of nm of the working electrode surface. Aogaki et al. reported that the field-induced forces acting near the electrode can in fact induce convection inside the diffusive layer [53] and influence the morphology of electrodeposits. For the Co/Pt film electrode in a multi-domain state, regarding it as a magnetized 2D sheet (Sect. 3), an average field of 0.2 T is expected at the film surface. As the current density involved is small, the average Lorentz force density at the surface is rather small, of the order of 10 N/m^3 ($j = 5$ mA/cm^2, $B_{av} = 0.2$ T), and cannot compete with the force driving natural convection ($\sim 10^3$ N/m^3). On the other hand, the magnetic field gradient force can act as a driving force. In our case, with the molar magnetic susceptibility $\chi_{mol} = 28.8 \times 10^{-9}$ m^3/mol for potassium ferricyanide of molar concentration of $c \approx 0.1$ M, we estimate the gradient force density to be of the order of 10^7 N/m^3 near the electrode surface. This force can inhibit the convection of paramagnetic species and can locally alter the concentration gradient. A study using CoPt nanowires embedded in an alumina membrane has shown that the oxygen reduction current can be enhanced by the magnetic field gradient driven convective inhibition of paramagnetic radicals near the electrode surface [54, 55].

9.6 Conclusions

Co/Pt multilayers with perpendicular magnetization are ideally suited for generating large localized magnetic field gradient forces while limiting the amplitude of the Lorentz force. Our multilayers produce values of ∇B^2 [equal to $(\boldsymbol{B}. \nabla)\boldsymbol{B}$] in excess of 10^6 T^2/m at the electrode surface. Use of a single Pt top layer as nonmagnetic cap layer allows us to convincingly compare the differences of behavior between magnetic and non-magnetic electrodes. For a reversible single e$^-$ electrochemical reaction, we show how the large magnetic field gradient force (Eq. 9.2) acting on the paramagnetic species impacts the diffusive double layer at the interface, decreasing its capacitance and its charge transfer resistance, indicating quantitatively how it perturbs the concentration profile near the interface. However, we find no indication of changes of kinetics of the reaction under the extreme local magnetic field gradient forces generated by our optimized nanoscale magnetic source.

We showed that our strategy for building high magnetic field gradient electrodes impacts the electrochemical process precisely where the forces act, specifically within the first tens of nm of electrolyte near the electrode surface. This strategy could be used beyond the presented benchmark case, for chemical processes highly sensitive to the electrode surface processes, such as the oxygen evolution reaction. Dynamical studies could possibly be extended down to very short time scales, limited only by the intrinsic properties of magnetic switching as the one could turn the magnetic field gradient forces on and off by magnetization reversal due to domain wall motion or spin–orbit torque [56, 57], possibly going down to the picosecond scale [58]. This should open new perspectives for impedance spectroscopy, where we can the define the timescale of a system's response as well as limiting the spatial location, with an excitation designed to occur only in the immediate vicinity of the working electrode.

Acknowledgements This project has received funding from the European Union's Horizon 2020 research and innovation programme under the Marie Skłodowska-Curie grant agreement No 766007. We also acknowledge the support of the University of Strasbourg, Institute for Advanced Studies (USIAS) and the Fondation Jean-Marie Lehn, as well as the support from IdEx Unistra (ANR 10 IDEX 0002), SFRI STRAT'US project (ANR 20 SFRI 0012) and EUR QMAT ANR-17-EURE-0024 under the framework of the French Investments for the Future Program.

References

1. J.M.D. Coey, Magnetoelectrochemistry. Europhys. News **34**(6), 246–248 (2003). https://doi.org/10.1051/epn:2003615
2. P. Dunne, Magnetochemistry and magnetic separation, in *Handbook of Magnetism and Magnetic Materials*, ed. by M. Coey, S. Parkin (Springer International Publishing, Cham, 2020), pp. 1–39. https://doi.org/10.1007/978-3-030-63101-7_35-1
3. Y. Zhang, C. Liang, J. Wu, H. Liu, B. Zhang, Z. Jiang et al., Recent advances in magnetic field-enhanced electrocatalysis. ACS Appl. Energy Mater. **3**(11), 10303–10316 (2020). https://doi.org/10.1021/acsaem.0c02104
4. F.A. Garcés-Pineda, M. Blasco-Ahicart, D. Nieto-Castro, N. López, J.R. Galán-Mascarós, Direct magnetic enhancement of electrocatalytic water oxidation in alkaline media. Nat. Energy **4**(6), 519–525 (2019). https://doi.org/10.1038/s41560-019-0404-4
5. P.C. Mondal, C. Fontanesi, D.H. Waldeck, R. Naaman, Field and chirality effects on electrochemical charge transfer rates: spin dependent electrochemistry. ACS Nano **9**(3), 3377–3384 (2015). https://doi.org/10.1021/acsnano.5b00832
6. P. Dunne, J.M.D. Coey, Influence of a magnetic field on the electrochemical double layer. J. Phys. Chem. C **123**(39), 24181–24192 (2019). https://doi.org/10.1021/acs.jpcc.9b07534
7. A.J. Bard, L.R. Faulkner, in *Electrochemical Methods: Fundamentals and Applications*, 2nd ed. (Wiley, 2000)
8. L.M.A. Monzon, J.M.D. Coey, Magnetic fields in electrochemistry: the Lorentz forcr. A mini-review. Electrochem. Comm. **42**, 38–41 (2014). https://doi.org/10.1016/j.elecom.2014.02.006
9. L.M.A. Monzon, J.M.D. Coey, Magnetic fields in electrochemistry: the Kelvin force. A mini-review. Electrochem. Comm. **42**, 42–45 (2014). https://doi.org/10.1016/j.elecom.2014.02.005
10. R. Aogaki, A. Tadano, K. Shinohara, MHD and Micro-MHD effects in electrochemical systems, in *Transfer Phenomena in Magnetohydrodynamic and Electroconducting Flows: Selected papers of the PAMIR Conference held in Aussois, France 22–26 September 1997*, ed. by A. Alemany, P. Marty, J.P. Thibault. Fluid Mechanics and Its Applications (Springer Netherlands, Dordrecht, 1999), pp. 169–179. https://doi.org/10.1007/978-94-011-4764-4_12
11. G. Hinds, F.E. Spada, J.M.D. Coey, T.R. Ní Mhíocháin, M.E.G. Lyons, Magnetic field effects on copper electrolysis. J. Phys. Chem. B **105**(39), 9487–9502 (2001). https://doi.org/10.1021/jp010581u
12. Y.-H. Li, Y.-J. Chen, The effect of magnetic field on the dynamics of gas bubbles in water electrolysis. Sci. Rep. **11**(1), 9346 (2021). https://doi.org/10.1038/s41598-021-87947-9
13. G. Mutschke, K. Tschulik, T. Weier, M. Uhlemann, A. Bund, J. Fröhlich, On the action of magnetic gradient forces in micro-structured copper deposition. Electrochim. Acta **55**(28), 9060–9066 (2010). https://doi.org/10.1016/j.electacta.2010.08.046
14. T.A. Butcher, J.M.D. Coey, Magnetic forces in paramagnetic fluids. J. Phys.: Condens. Mat. **35**(5), 053002. https://doi.org/10.1088/1361-648X/aca37f
15. P.A. Dunne, J. Hilton, J.M.D. Coey, Levitation in paramagnetic liquids. J. Magn. Magn. Mater. **316**(2), 273–276 (2007). https://doi.org/10.1016/j.jmmm.2007.02.128
16. D. Fernández, Z. Diao, P. Dunne, J.M.D. Coey, Influence of magnetic field on hydrogen reduction and co-reduction in the Cu/CuSO4 system. Electrochim. Acta **55**(28), 8664–8672 (2010). https://doi.org/10.1016/j.electacta.2010.08.004
17. P. Dunne, L. Mazza, J.M.D. Coey, Magnetic structuring of electrodeposits. Phys. Rev. Lett. **107**(2), 024501 (2011). https://doi.org/10.1103/PhysRevLett.107.024501
18. P. Dunne, R. Soucaille, K. Ackland, J.M.D. Coey, Structuring of electrodeposits with permanent magnet arrays. Magnetohydrodynamics **48**, 331–341 (2012)
19. P. Dunne, J.M.D. Coey, Patterning metallic electrodeposits with magnet arrays. Phys. Rev. B **85**(22), 224411 (2012). https://doi.org/10.1103/PhysRevB.85.224411
20. J.M.D. Coey, R. Aogaki, F. Byrne, P. Stamenov, Magnetic stabilization and vorticity in submillimeter paramagnetic liquid tubes. Proc. Natl. Acad. Sci. **106**(22), 8811–8817 (2009). https://doi.org/10.1073/pnas.0900561106

21. P. Dunne, T. Adachi, A.A. Dev, A. Sorrenti, L. Giacchetti, A. Bonnin et al., Liquid flow and control without solid walls. Nature **581**(7806), 58–62 (2020). https://doi.org/10.1038/s41586-020-2254-4

22. S. Wang, J. Zhang, O. Gharbi, V. Vivier, M. Gao, M.E. Orazem, Electrochemical impedance spectroscopy. Nat. Rev. Methods Prim. **1**(1), 1–21 (2021). https://doi.org/10.1038/s43586-021-00039-w

23. C.P. de Abreu, C.M. de Assis, P.H. Suegama, I. Costa, M. Keddam, H.G. de Melo et al., Influence of probe size for local electrochemical impedance measurements. Electrochim. Acta **233**, 256–261 (2017). https://doi.org/10.1016/j.electacta.2017.03.017

24. A.-T. Tran, F. Huet, K. Ngo, P. Rousseau, Artefacts in electrochemical impedance measurement in electrolytic solutions due to the reference electrode. Electrochim. Acta **56**(23), 8034–8039 (2011). https://doi.org/10.1016/j.electacta.2010.12.088

25. G.J. Brug, A.L.G. van den Eeden, M. Sluyters-Rehbach, J.H. Sluyters, The analysis of electrode impedances complicated by the presence of a constant phase element. J. Electroanal. Chem. Interfacial Electrochem. **176**(1), 275–295 (1984). https://doi.org/10.1016/S0022-0728(84)80324-1

26. B. Hirschorn, M.E. Orazem, B. Tribollet, V. Vivier, I. Frateur, M. Musiani, Determination of effective capacitance and film thickness from constant-phase-element parameters. Electrochim. Acta **55**(21), 6218–6227 (2010). https://doi.org/10.1016/j.electacta.2009.10.065

27. L. Fallarino, A. Oelschlägel, J.A. Arregi, A. Bashkatov, F. Samad, B. Böhm et al., Control of domain structure and magnetization reversal in thick Co/Pt multilayers. Phys. Rev. B **99**(2), 024431 (2019). https://doi.org/10.1103/PhysRevB.99.024431

28. O. Hellwig, A. Berger, J.B. Kortright, E.E. Fullerton, Domain structure and magnetization reversal of antiferromagnetically coupled perpendicular anisotropy films. J. Magn. Magn. Mater. **319**(1), 13–55 (2007). https://doi.org/10.1016/j.jmmm.2007.04.035

29. A. Hubert, R. Schäfer, Domain theory, in *Magnetic Domains: The Analysis of Magnetic Microstructures*, ed. A. Hubert, R. Schäfer (Springer, Berlin, Heidelberg, 1998), pp. 99–335. https://doi.org/10.1007/978-3-540-85054-0_3

30. C. Kittel, Theory of the structure of ferromagnetic domains in films and small particles. Phys. Rev. **70**(11–12), 965–971 (1946). https://doi.org/10.1103/PhysRev.70.965

31. C. Kittel, Physical theory of ferromagnetic domains. Rev. Mod. Phys. **21**(4), 541–583 (1949). https://doi.org/10.1103/RevModPhys.21.541

32. C. Kooy, U. Enz, Experimental and theoretical study of the domain configuration in thin layers of BaFe12O19. Philips Res. Rep. **7**(29) (1960)

33. Z. Diao, E.R. Nowak, G. Feng, J.M.D. Coey, Magnetic noise in structured hard magnets. Phys. Rev. Lett. **104**(4), 047202 (2010). https://doi.org/10.1103/PhysRevLett.104.047202

34. V. Kamberský, P. de Haan, J. Šimšová, S. Porthun, R. Gemperle, J.C. Lodder, Domain wall theory and exchange stiffness in Co/Pd multilayers. J. Magn. Magn. Mater. **157–158**, 301–302 (1996). https://doi.org/10.1016/0304-8853(95)01162-5

35. O. Hellwig, G.P. Denbeaux, J.B. Kortright, E.E. Fullerton, X-ray studies of aligned magnetic stripe domains in perpendicular multilayers. Physica B **336**(1), 136–144 (2003). https://doi.org/10.1016/S0921-4526(03)00282-5

36. R. Salikhov, F. Samad, B. Böhm, S. Schneider, D. Pohl, B. Rellinghaus et al., Control of Stripe-Domain-Wall Magnetization in Multilayers Featuring Perpendicular Magnetic Anisotropy. Phys. Rev. Appl. **16**(3), 034016 (2021). https://doi.org/10.1103/PhysRevApplied.16.034016

37. E. Furlani, *Permanent Magnet and Electromechanical Devices: Materials, Analysis, and Applications*, 1st ed. (Academic Press, 2001)

38. P. Dunne, pdunne/pymagnet: Pymagnet—Fields, Forces and Torques (2021)

39. P.H. Daum, C.G. Enke, Electrochemical kinetics of the ferri-ferrocyanide couple on platinum. Anal. Chem. **41**(4), 653–656 (1969). https://doi.org/10.1021/ac60273a007

40. N.P.C. Stevens, M.B. Rooney, A.M. Bond, S.W. Feldberg, A comparison of simulated and experimental voltammograms obtained for the [Fe(CN)6]3-/4—couple in the absence of added supporting electrolyte at a rotating disk electrode. J. Phys. Chem. A **105**(40), 9085–9093 (2001). https://doi.org/10.1021/jp0103878

41. H. Liu, L. Pan, H. Huang, Q. Qin, P. Li, J. Wen, Hydrogen bubble growth at micro-electrode under magnetic field. J. Electroanal. Chem. **754**, 22–29 (2015). https://doi.org/10.1016/j.jelechem.2015.06.015

42. C. Beriet, D. Pletcher, A microelectrode study of the mechanism and kinetics of the ferro/ferricyanide couple in aqueous media: the influence of the electrolyte and its concentration. J. Electroanal. Chem. **361**(1), 93–101 (1993). https://doi.org/10.1016/0022-0728(93)87042-T

43. L.M. Fischer, M. Tenje, A.R. Heiskanen, N. Masuda, J. Castillo, A. Bentien et al., Gold cleaning methods for electrochemical detection applications. Microelectron. Eng. **86**(4), 1282–1285 (2009). https://doi.org/10.1016/j.mee.2008.11.045

44. X. Gao, J. Lee, H.S. White, Natural convection at microelectrodes. Anal. Chem. **67**(9), 1541–1545 (1995). https://doi.org/10.1021/ac00105a011

45. J. Moldenhauer, M. Meier, D.W. Paul, Rapid and direct determination of diffusion coefficients using microelectrode arrays. J. Electrochem. Soc. **163**(8), H672 (2016). https://doi.org/10.1149/2.0561608jes

46. S.J. Konopka, B. McDuffie, Diffusion coefficients of ferri- and ferrocyanide ions in aqueous media, using twin-electrode thin-layer electrochemistry. Anal. Chem. **42**(14), 1741–1746. https://doi.org/10.1021/ac50160a042

47. R.S. Nicholson, Theory and application of cyclic voltammetry for measurement of electrode reaction kinetics. Anal. Chem. **37**(11), 1351–1355 (1965). https://doi.org/10.1021/ac60230a016

48. M. Stieble, K. Jüttner, Surface blocking in the redox system $Pt/[Fe(CN)6]^{3-},[Fe(CN)6]^{4-}$: an ac impedance study. J. Electroanal. Chem. Interfacial Electrochem. **290**(1), 163–180 (1990). https://doi.org/10.1016/0022-0728(90)87428-M

49. D.A.C. Brownson, C.E. Banks, *The Handbook of Graphene Electrochemistry* (Springer, London, 2014). https://doi.org/10.1007/978-1-4471-6428-9

50. G.P. Rao, S.K. Rangarajan, A new relaxation method for studying electrode reactions. J. Electroanal. Chem. Interfacial Electrochem. **41**(3), 473–489 (1973). https://doi.org/10.1016/S0022-0728(73)80425-5

51. R. Aogaki, K. Fueki, T. Mukaibo, Diffusion process in viscous flow of electrolyte solution in magnetohydrodynamic pump electrodes. Denki Kagaku oyobi Kogyo Butsuri Kagaku **44**(2), 89–94 (1976). https://doi.org/10.5796/kogyobutsurikagaku.44.89

52. O. Devos, O. Aaboubi, J.-P. Chopart, A. Olivier, C. Gabrielli, B. Tribollet, Is there a magnetic field effect on electrochemical kinetics? J. Phys. Chem. A **104**(7), 1544–1548 (2000). https://doi.org/10.1021/jp993696v

53. R. Aogaki, R. Morimoto, Nonequilibrium fluctuations in micro-MHD effects on electrodeposition. IntechOpen (2011). https://doi.org/10.5772/21230

54. N.B. Chaure, F.M.F. Rhen, J. Hilton, J.M.D. Coey, Design and application of a magnetic field gradient electrode. Electrochem. Commun. **9**(1), 155–158 (2007). https://doi.org/10.1016/j.elecom.2006.08.059

55. N.B. Chaure, J.M.D. Coey, Enhanced oxygen reduction at composite electrodes producing a large magnetic gradient. J. Electrochem. Soc. **156**(3), F39 (2009). https://doi.org/10.1149/1.3000576

56. P. Dunne, C. Fowley, G. Hlawacek, J. Kurian, G. Atcheson, S. Colis et al., Helium ion microscopy for reduced spin orbit torque switching currents. Nano Lett. **20**(10), 7036–7042 (2020). https://doi.org/10.1021/acs.nanolett.0c02060

57. J. Kurian, A. Joseph, S. Cherifi-Hertel, C. Fowley, G. Hlawacek, P. Dunne et al., Deterministic multi-level spin orbit torque switching using focused He$^+$ ion beam irradiation. Appl. Phys. Lett. **122**(3), 032402 (2023). https://doi.org/10.1063/5.0131188

58. K. Jhuria, J. Hohlfeld, A. Pattabi, E. Martin, A.Y. Arriola Córdova, X. Shi et al., Spin–orbit torque switching of a ferromagnet with picosecond electrical pulses. Nat. Elect. **3**(11), 680–686 (2020). https://doi.org/10.1038/s41928-020-00488-3

Chapter 10
Interferometric Measurement of Forward Reaction Rate Order and Rate Constant of a Dy(III)-PC88A-HCl Solvent Extraction System

Fengzhi Sun, Kilian Ortmann, Kerstin Eckert, and Zhe Lei

10.1 Introduction

Rare-earth elements (REEs) are indispensable for numerous high-tech products, such as motors for electric cars and permanent magnets for generators [1]. As energy production shifts to a higher proportion of renewable sources, the demand for REEs is foreseen to increase greatly over the next few decades [2]. However, the beneficiation of REEs involves the environmentally and economically costly Solvent eXtraction (SX) process. Ever since its introduction in the 1960s, this continues to be a low-efficiency procedure on the industrial scale, limited by the small separation factor and high consumption of non-recyclable chemicals [3]. Furthermore, there is increased interest in a stabler REE supply chain with an additional focus on innovative separation methods that are more efficient in terms of both REE beneficiation and secondary resources, e.g., recycling from end-of-life REEs [4]. In recent years, reports on the working principle of enriched REE ions close to their magnetic source

F. Sun · K. Ortmann · K. Eckert · Z. Lei (✉)
Institute of Fluid Dynamics, Helmholtz-Zentrum Dresden-Rossendorf (HZDR), 01328 Dresden, Germany
e-mail: z.lei@hzdr.de

K. Ortmann
e-mail: k.ortmann@hzdr.de

K. Eckert
e-mail: k.eckert@hzdr.de

K. Eckert · Z. Lei
Institute of Processing Engineering and Environmental Technology, Technische Universität Dresden, 01069 Dresden, Germany

B. Doudin et al. (eds.), *Magnetic Microhydrodynamics*, Topics in Applied Physics 120,
https://doi.org/10.1007/978-3-031-58376-6_10

[5–8] have raised the prospect of green REE separation technology, which should be fully explored. The stand-alone technology can also be combined with conventional SX to produce an innovative magnet-assisted solvent extraction process. By addressing the differences in REEs' magnetic susceptibility, an intrinsic property of the respective metals [8], the selective enhancement of individual REEs' extraction kinetics is expected. An enhanced separation factor is expected to be achieved by sampling prior to equilibrium, as is done nowadays in SX plants. However, to elucidate the mechanism for selectively enhancing the extraction kinetics in magnet-assisted solvent extraction, a novel approach for investigating the kinetic mechanism is crucial.

The mixer-settler is the main device within which solvent extraction takes place in a separation plant. Thus, hundreds of mixer-settler stages linked in a countercurrent circuit might be needed to achieve sufficient separation quality for individual REE products [3]. The phosphorous acids di-2-ethylhexylphosphoric acid (D2EHPA) and 2-ethylhexylphosphonic acid mono-2-ethylhexyl ester (PC88A) are two widely used cation exchanger extractants [9]. They tend to form dimers, i.e., $H_2A_{2,O}$, in a non-polar organic phase. The extraction reaction equation for Dy reads

$$Dy_A^{3+} + 3H_2A_{2,O} \leftrightarrow Dy(HA_2)_{3,O} + 3H_A^+. \tag{10.1}$$

Knowledge of the distribution coefficient

$$D_{RE} = \frac{c_{RE(o)}}{c_{RE(a)}} \tag{10.2}$$

is defined by the concentration ratio of REEs, i.e., Dy in Eq. (10.1), in the organic phase and that in the aqueous phase at equilibrium; and the separation factor

$$\beta_{RE1/RE2} = \frac{D_{RE1}}{D_{RE2}} \tag{10.3}$$

which is the ratio between the distribution coefficient, respectively, for two RE species, namely *RE1* and *RE2* in subscript. The $c_{RE,o}$ and $c_{RE,a}$ are REE concentrations in the organic and aqueous phases at equilibrium, respectively, and are critical in designing the process. This information could further feed into equilibrium-based modeling to improve and optimize the process [10]. Furthermore, the kinetic mechanism offers critical information about the timescale and hence the size of the plant for the separation process. Nevertheless, few tools have been developed to investigate chemical changes occurring at the liquid–liquid interface. The exploration of the kinetics is mainly based on indirect experimental investigations. Several sets of apparatus have been used to study the initial boundary mass flux where the concentration is known and the partial rate order for individual species and the reaction rate constant can be individually studied. Since only the species at the initial time step have defined concentration values, the forward rate equation normally takes a semi-empirical form

$$R\left[\frac{mol}{m^2 \cdot s}\right] = k_f c^a_{RE(a)} c^b_{H_2A_2(o)} c^c_{H+}. \tag{10.4}$$

The Lewis cell and its variants represent a conventional experimental approach to quantifying the solvent extraction kinetics of this heterogeneous extraction process [11]. The aqueous phase and organic phase with their respectively defined volumes, V, are stirred separately/spontaneously so that respective bulk phases are well mixed while the water–oil interface is still stratified. The average extracted rare-earth $\overline{c}_{RE(a)}$ is quantified by characterizing the sampled species at different time intervals during the reaction. Consequently, the reaction rate R of RE(III) is readily computed with

$$R\left[\frac{mol}{m^2 \cdot s}\right] = \frac{V}{A} \cdot \frac{d\overline{c}_{RE(a)}}{dt} \tag{10.5}$$

and the initial rate is then extrapolated at the zero time stamp. There are some obvious disadvantages to Lewis-type cells. Tracking the reaction at different time intervals requires independent, repetitive experiments. It further entails the need to independently study the influence of the REE concentration, extractant concentration and pH. The overall sample consumption is enormous (100–1000 mL per experiment). Plus, at an increased stirrer speed, the hydrodynamic situation is ill-defined at the interface. Kelvin–Helmholtz instability and entrainment can occur at interfaces with an increased shear rate [12]. Consequently, a complicated wave pattern, turbulence and emulsification occur and the initially well-defined interface area A changes. Using a Lewis-type cell with a membrane, Chitra et al. [13] found a first-order rate law for Nd(III) and a third-order law for a dimeric PC88A concentration for the forward reaction, while the backward reaction is of the first order for the REE complex and of the third order for H+. The reaction order is, surprisingly, the same as the stoichiometric coefficient of the reactants in Eq. (10.1). Mishra found a quasi-first order rate for La(III) and D2EHPA and 0.08 order for H+ in the presence of lactic acid [14] for the forward reaction. Alternatively, using highly stirred apparatus AKUFVE [15], continuous sampling at different time intervals is achieved by sacrificing the quantified interface area. Wang et al. [16] found a reaction order of 1 for Sm(III), Eu(III), Gd(III) and PC88A, and a reaction order of -1 for H+ at a stir speed of 420 rpm. At the same time, there was one attempt [17] to minimize the organic phase volume in a single drop experiment where the organic phase rises freely due to buoyancy in a REE-laden aqueous column. The extraction time is adjusted by changing the column height, leading the setup to deliver more flexible measurements, with the advantage of sampling at different time intervals during the reaction in a single run. Mostaedi et al. [17] found a forward reaction law of the first order for Sm(III) and D2EHPA and one of -0.8 for H+, while Huang [18] found a quasi-first order rate for La(III), Gd(III), Ho(III) and Lu(III) concentrations and a PC88A monomeric concentration and quasi-negative first order for H+. An even smaller amount of sample is needed when employing the hollow-fiber membrane extractor technique, using which Nakashio et al. [19] reported very different rate laws compared with other studies.

The increased effective diffusivity might cause an over-estimated mass flux due to Taylor diffusion [20] as a result of shear flow in the porous medium.

One major reason for the current disagreements over the rate law is un-decoupled fluid dynamics in the investigation apparatus. One obvious dilemma lies in the need to stir the cell to achieve active mixing in the accurate sampling that enters $\bar{c}_{RE(a)}$ in Eq. (10.5) and the clearly defined reaction area when the interface is free from convection. In this paper, we propose and deliver preliminary validation for using a Mach–Zehnder interferometer to quantify the rate law with minimum reagent consumption, in the order of 1 mL per experiment. The heterogeneous extraction process, similar to a Lewis cell without active stirring, takes place at the water–oil interface, with both bulk phases in a state of stagnancy. Therefore, a space- and time-resolved Dy(III) boundary layer is monitored in the vicinity of the interface, where cation exchange takes place with PC88A as an extractant. A diffusion layer of Dy(III) is formed gradually with a positive concentration gradient in the direction of gravity. Further integrated spatially, a Dy(III) boundary flux is quantified from which the forward reaction rate order and rate constant are studied by parametrically varying the initial Dy(III), H_2A_2 concentration.

10.2 Experimental Aspects

This experiment addresses solvent extraction in an immiscible water–oil system. For that purpose, $DyCl_3$ up to 1 M (purity 99.9%, Abcr GmbH) with the pH adjusted to 1 by HCl in an aqueous phase is brought into contact with an organic phase, a low-viscosity paraffin (Sigma-Aldrich 76,235), dissolving the extractant PC88A up to 1 M. Hence, a Dy(III) cation exchange [3] following Eq. (10.1) is triggered at the water–oil interface. The reaction is conducted using a quartz-glass cell (Hellma 404-1) with inner dimensions of $36.5 \times 18.5 \times 1$ mm^3. The narrow gap width of the cell allows us to study the diffusion-extraction kinetics process in a quasi-2D Hele-Shaw configuration (Fig. 10.1a), a widely used fluid mechanics model setup for flow visualization and qualification, incorporating advective transport [21, 22]. The cell is then placed in the measurement arm of a 632.8 nm monochromatic laser-based Mach–Zehnder interferometer [5–7], with the cell's depth in the direction of the laser beam. The refractive index of the aqueous phase, which is proportional to the RE(III) concentration change, is monitored by the phase shift of the interference fringes at a frame rate of 5 fps with a resolution of 2160×2160 pixels (Jai Go-5100 M-USB). The interferogram, see Fig. 10.1b, is processed [5–7] into a space- and time-resolved RE(III) concentration field as shown in Fig. 10.2 in the following chapter.

To prepare the cell, the aqueous phase is first poured into the bottom until a fixed height of approximately half the cell is reached, where we set the interface coordinate at $z = 0$ mm, see Fig. 10.1b. A fixed camera view is maintained so that the measurement region at any parametric variation observes the same respective area in the aqueous phase, with the water–oil interface acting as a boundary. Then, after the aqueous phase stabilizes, an equal amount of organic phase is injected

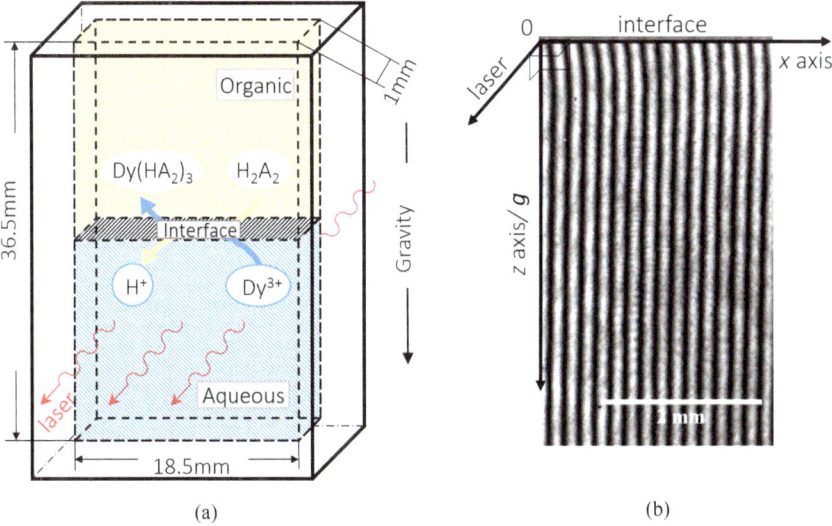

Fig. 10.1 Illustration of **a** experimental configuration. The coordinate system is selected so that gravity is parallel to the z direction with the water–oil interface at $z = 0$ mm. **b** interferogram at aqueous phase with water–oil interface as upper boundary

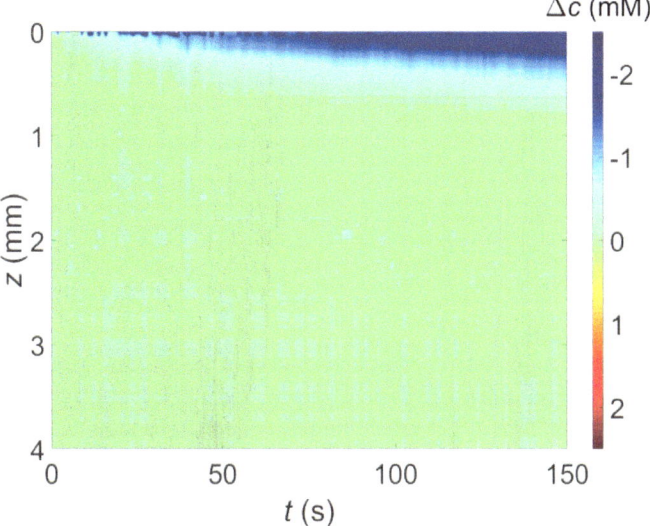

Fig. 10.2 Dy(III) concentration distribution along z axis against different time intervals after the reaction layer with an initial concentration of DyCl$_3$ 1 M and PC88A 0.25 M

simultaneously from both ports at the top of the cell. Using this method, the two fronts of organic phase meet in the middle of the cell and the interface remains relatively flat. Menisci near the cell edge can be observed when the two phases are in contact (see Fig. 10.1a). As a result of RE(III) consumption, the density gradient is parallel to gravity. Therefore, the hydrodynamically stable system, together with a flat interface in the middle of the cell, further simplifies the model into a quasi-1-D diffusion reaction transport [5–7].

10.3 Results

Figure 10.2 shows an experimentally resolved Dy(III) mass boundary layer in the aqueous phase over time, obtained by interferometry, see Fig. 10.1b. The diffusion front scales with a $t^{1/2}$ law, which indicates a diffusive transport process of Dy(III) concentration stratifications. The quasi-1D reaction–diffusion scenario makes it possible to de-noise the experimental illustrations by averaging the concentration profile along the x axis, i.e., in the direction of the interface (see Fig. 10.1b), for every z-axis value, i.e., in the direction of gravity. The interface position is located at $z = 0$ and the Dy(III) cation exchange starts at time $t = 0$ s. The experimental conditions of Fig. 10.2b in this case are a DyCl$_3$ and PC88A with a molarity of, respectively, 0.25 M and 1 M in their respective phases. The concentration profile is shown as the reduction in concentration (Δc) versus the distance (z axis) from the interface at selected times t.

To this end, we revisit Eqs. (10.4) and (10.5) to explicitly correlate the rate law and the results gained from the interferometer. Under the condition that the reverse reaction is negligible,

$$
\log_{10} \frac{V}{A} \cdot \frac{d\bar{c}_{RE(a)}}{dt} \\
= \log_{10} k_f + a \log_{10} c_{RE(a)} + b \log_{10} c_{H_2A_2(o)} + \log_{10} c_{H+}
\tag{10.6}
$$

the interferometer results (left) can be correlated with the rate law (right) by applying the log operator on both sides. In this way, varying the individual species, i.e., Dy(III) and PC88A concentration or pH value, the rate order can be individually extracted.

10.3.1 Dy(III) Concentration Influence

The Dy(III) concentration is varied in an aqueous phase with a salinity of 0.25, 0.5, 0.75 and 1 M while the extractant concentration remains the same, i.e., 0.25 M PC88A in the organic phase. The result, shown in Fig. 10.2, is then integrated along the z axis so that the value $V/A \cdot \bar{c}_{RE(a)}$ is obtained at different time intervals during the reaction, see Fig. 10.3a. A linear fit (dashed line in Fig. 10.3a) is then applied to individual

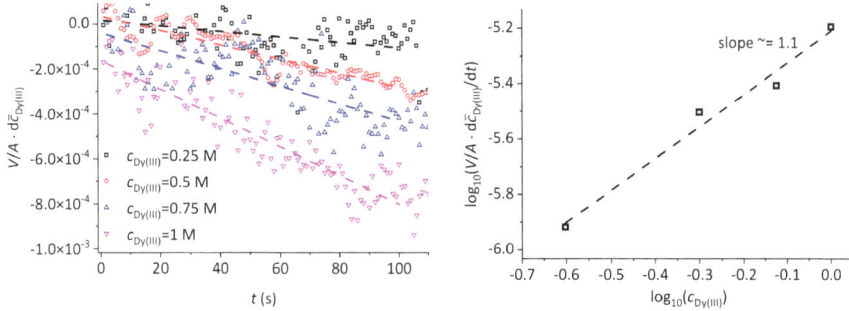

Fig. 10.3 The interface-averaged spatial integral of Dy(III) concentration, $V/A \cdot \bar{c}_{RE(a)}$, with PC88A 0.25 M, i.e., dimeric concentration 0.125 M, at different initial Dy(III) concentrations (**a**). The dashed lines are linear fits for different initial Dy(III) concentrations; their slope is summarized in **b**. **b** initial forward reaction rate at different initial Dy(III) concentrations in a double logarithmic diagram

parametric variations with its gradient feeding to the left-hand side of Eq. (10.6). Summarizing the $V/A \cdot d\bar{c}_{RE(a)}/dt$ value against different Dy(III) concentrations in a double logarithmic diagram (Fig. 10.3b), a quasi-first order dependence of Dy(III) is found for experimental system.

10.3.2   Influence of Dimeric PC88A Concentration

The extractant PC88A(III) concentration is varied in organic phases of 0.25, 0.5, 0.75 and 1 M while the Dy(III) concentration remains the same, i.e., 0.25 M DyCl$_3$ in the aqueous phase. The resulting dimeric PC88A concentration is half of the respective values. The result, shown in Fig. 10.2, is then integrated along the z axis so that the value $V/A \cdot \bar{c}_{RE(a)}$ is obtained at different time intervals during the reaction, see Fig. 10.4a. A linear fit (dashed line in Fig. 10.4a) is then applied to individual parametric variations with its time derivative feeding to the left-hand side of Eq. (10.6). Summarizing the $V/A \cdot d\bar{c}_{RE(a)}/dt$ value against different PC88A concentrations in a $\log_{10}$-$\log_{10}$ diagram, Fig. 10.4a, b quasi-second order dependence of Dy(III) is found.

Consequently, the rate law is readily computed by combining the intercept and slope of the linear fit in Figs. 10.3b and 10.4b, respectively. To this end, the forward reaction rate takes a fixed pH value, hence $k_{f'} = k_f c_{H+}^c = 7.77 \times 10^{-4}$

$$R\left[\frac{mol}{m^2 \cdot s}\right] = 7.77 \times 10^{-4}\left[\frac{m}{M^{-2.4} \cdot s}\right] c_{RE(a)}^{1.1} c_{H_2 A_2(o)}^{2.3}$$

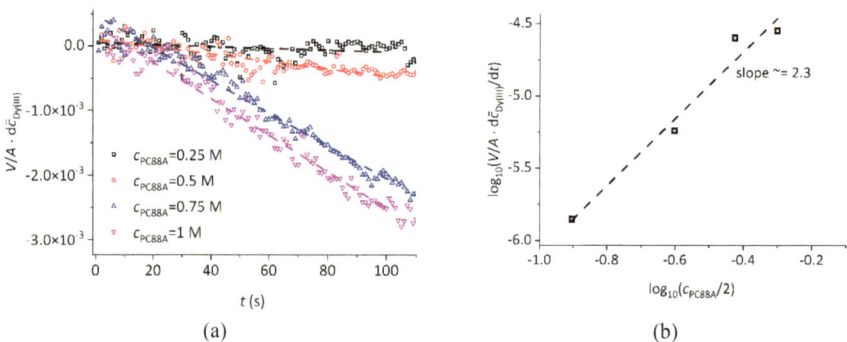

Fig. 10.4 The interface-averaged spatial integral of the Dy(III) concentration, $V/A \cdot \overline{c}_{RE(a)}$, with Dy(III) 0.25 M at different initial PC88A concentrations of 0.25 M, 0.5 M and 1 M, i.e., dimeric concentrations of 0.125 M 0.125 M, 0.25 M and 0.5 M, respectively (**a**). The dashed lines are linear fits for different initial PC88A concentrations, with their slopes summarized in **b**. **b** the initial forward reaction rate at different initial dimeric PC88A concentrations in a double logarithmic diagram

10.4 Conclusion

We report a novel approach for measuring the forward rate law of a rare earth (RE) solvent extraction system. A cation exchange system of Dy(III)-PC88A-HCl is used for validation. A Mach–Zehnder interferometer monitors the space- and time-resolved Dy(III) concentration boundary layer in a Hele-Shaw cell. The dependence of both the Dy(III) and the dimeric extractant PC88A concentration are studied independently and a rate law is found following a quasi-first order and quasi-second order, respectively. Our method is superior to conventional methods as it requires 2–3 orders of magnitude less material, and Dy(III) can be traced at different time intervals after the reaction in one experiment without the probe disturbance that is an issue with conventional methods. In addition, the approach can genuinely be applied to all kinetic systems, provided the medium is transparent or translucent. The potentially interesting aspects to be followed up include the extraction kinetic law for all trivalent RE ions, not limited to the model system of the Dy(III)-PC88A-HCl solvent extraction system we validated here. While an active follow-up investigation is underway extending the parametric range and systems to refine the accuracy and expand the rate law to encompass a broader spectrum of reaction constant determination, this work focuses on establishing and validating the general method.

Acknowledgements We are grateful for the financial support by the German Aerospace Center (DLR) with funds provided by the Federal Ministry for Economic Affairs and Energy (BMWi) based on an act of the German Bundestag under Grant DLR 50WM2059 (Project MAGSOLEX). We thank our colleague J. Heinrich for the interesting discussion on chemical reaction kinetic experimentation.

References

1. K.M. Goodenough, F. Wall, D. Merriman, The rare earth elements: demand, global resources, and challenges for resourcing future generations. Nat. Resour. Res. **27**, 201–216 (2018)
2. L. Baldi, M. Peri, D. Vandone, Clean energy industries and rare earth materials: economic and financial issues. Energy Policy **66**, 53–61 (2014)
3. F. Xie, T.A. Zhang, D. Dreisinger, F. Doyle, A critical review on solvent extraction of rare earths from aqueous solutions. Miner. Eng. **56**, 10–28 (2014)
4. K. Binnemans, Y. Pontikes, P. T. Jones, T. V. Gerven, B. Blanpain, Recovery of rare earths from industrial waste residues: a concise review. *Proceedings of the 3rd international slag valorisation symposium: the transition to sustainable materials management* Slag Valorisation Symposium, pp. 191–205 (2013)
5. Z. Lei, B. Fritzsche, K. Eckert, Evaporation-assisted magnetic separation of rare-earth ions in aqueous solutions. J. Phys. Chem. C **121**(44), 24576–24587 (2017)
6. Z. Lei, B. Fritzsche, K. Eckert, Magnetic separation of rare-earth ions: transport processes and pattern formation. Phys. Rev. Fluids **6**(2), L021901 (2021)
7. Z. Lei, B. Fritzsche, K. Eckert, Stability criterion for the magnetic separation of rare-earth ions. Phys. Rev. E **101**(1), 013109 (2020)
8. Z. Lei, B. Fritzsche, R. Salikhov, K. Schwarzenberger, O. Hellwig, K. Eckert, Magnetic separation of rare-earth ions: property database and kelvin force distribution. J. Phys. Chemistry C **126**(4), 2226–2233 (2022)
9. N.A. Ismail, A. Nurul, M.Y. Yunus, A. Hisyam, Selection of extractant in rare earth solvent extraction system: a review. Int. J. Recent Tech. Eng. **8**(1), 728–743 (2019)
10. T. Zhu, Solvent extraction in China. Hydrometallurgy **27**(2), 231–245 (1991)
11. S. Yin, S. Li, B. Zhang, J. Peng, L. Zhang, Extraction kinetics of neodymium (III) from chloride medium in the presence of two complexing agents by D2EHPA using a constant interfacial area cell with laminar flow. Hydrometallurgy **161**, 160–165 (2016)
12. P.G. Drazin, Kelvin-Helmholtz instability of finite amplitude. J. Fluid Mech. **42**(2), 321–335 (1970)
13. K.R. Chitra, A.G. Gaikwad, G.D. Surender, A.D. Damodaran, Studies on kinetics of forward and backward extraction of neodymium by using phosphonic acid monoester as an acidic extractant. Chem. Eng. J. Biochem. Eng. J. **60**(1–3), 63–73 (1995)
14. S. Satpathy, S. Mishra, Kinetics and mechanisms of solvent extraction and separation of La (III) and Ni (II) with DEHPA in petrofin. Trans. Nonferrous Metals Soc. China **29**(7), 1538–1548 (2019)
15. G.W. Stevens, M.P. Jilska, Kinetics of solvent extraction processes. Miner. Process. Extr. Metall. Rev. **17**(1–4), 205–226 (1997)
16. C. Zhuo, Y. Wang, Solvent extraction kinetics of Sm (III), Eu (III) and Gd (III) with 2-ethylhexyl phosphoric acid-2-ethylhexyl ester. Chin. J. Chem. Eng. **26**(2), 317–321 (2018)
17. R. Torkaman, J. Safdari, M. Torab-Mostaedi, M.A. Moosavian, A kinetic study on solvent extraction of samarium from nitrate solution with D2EHPA and Cyanex 301 by the single drop technique. Hydrometallurgy **150**, 123–129 (2014)
18. W. Cao, K. Huang, X. Wang, H. Liu, Extraction kinetics and kinetic separation of La (III), Gd (III), Ho (III) and Lu (III) from chloride medium by HEHEHP. J. Rare Earths **39**(10), 1264–1272 (2021)
19. F. Kubota, M. Goto, F. Nakashio, T. Hano, Extraction kinetics of rare earth metals with 2-ethylhexyl phosphonic acid mono-2-ethylhexyl ester using a hollow fiber membrane extractor. Sep. Sci. Technol. **30**(5), 777–792 (1995)
20. B.M. Alessio, S. Shim, A. Gupta, H.A. Stone, Diffusioosmosis-driven dispersion of colloids: a Taylor dispersion analysis with experimental validation. J. Fluid Mech. **942**, A23 (2022)
21. P. Gondret, M. Rabaud, Shear instability of two-fluid parallel flow in a Hele-Shaw cell. Phys. Fluids **9**(11), 3267–3274 (1997)
22. Y. Shi, K. Eckert, A novel Hele-Shaw cell design for the analysis of hydrodynamic instabilities in liquid–liquid systems. Chem. Eng. Sci. **63**, 3560–3563 (2008)

Part IV
Applications in Biology

Chapter 11
Magnetoelastic Elastomers and Hydrogels for Studies of Mechanobiology

Peter A. Galie, Katarzyna Pogoda, Kiet A. Tran, Andrejs Cēbers, and Paul A. Janmey

11.1 Introduction

The mechanical properties of materials to which biological cells adhere are now commonly accepted to be a major factor in determining cell function, proliferation, and differentiation [1]. These mechanical properties, usually quantified by elastic modulus, can change either gradually or rapidly during processes such as tissue differentiation, development of fibrotic disease, changes in vascular pressure as the heart beats, sudden impact, or the effects of gravity during activities such as sitting or walking. Many efforts have been directed at developing biologically compatible materials with elastic moduli similar to those of biological tissues [2], and in particular development of methods by which the elastic modulus of a cell substrate can be changed to study how cells respond to this change in substrate stiffness [3]. Most such efforts involve chemical strategies to break network strands or introduce or eliminate network crosslinks, which can change the elastic modulus by several factors over a period of minutes to hours [4, 5]. Such strategies mimic some aspects of normal or pathological stiffness changes but are limited by the relatively slow rate

P. A. Galie · K. A. Tran
Department of Biomedical Engineering, Rowan University, Glassboro, NJ, USA

K. Pogoda
Department of Experimental Physics of Complex Systems, Institute of Nuclear Physics PAN, Krakow, Poland

A. Cēbers
Department of Physics, University of Latvia, Riga, Latvia

P. A. Janmey (✉)
Departments of Physiology and Physics & Astronomy, University of Pennsylvania, Philadelphia, PA, USA
e-mail: janmey@pennmedicine.upenn.edu

© The Author(s) 2024

B. Doudin et al. (eds.), *Magnetic Microhydrodynamics*, Topics in Applied Physics 120, https://doi.org/10.1007/978-3-031-58376-6_11

at which the elastic modulus changes and usually by irreversibility of the chemically induced stiffness change. To reproduce the rapid and reversible stiffness changes that occur as the heart beats, blood vessels pulse, or soft tissues are deformed by muscle contraction, chemically induced stiffening or softening materials are inadequate.

An alternative method to change the stiffness of a soft elastomer or a hydrogel is the introduction of ferromagnetic particles into the material and then subjecting the composite material to magnetic fields [6–10]. The most commonly used such materials are magnetoelastomers in which particles such as carbonyl iron spheres are embedded in polydimethylsiloxane (PDMS) or other similar soft materials that have elastic moduli similar to those of some mammalian tissues. These materials have been extensively studied and quantitatively analyzed, with excellent fits of theory to experiment [11, 12]. In part, understanding the effect of magnetic fields on material stiffness is facilitated by the fact that the host rubber-like material, such as PDMS, is linearly elastic, with shear or Youngs moduli that are nearly independent of frequency or strain magnitude over the range of timescales and deformation magnitudes that occur in vivo. Similar considerations also apply to polyacrylamide or other flexible polymer hydrogels embedded with ferromagnetic particles [6, 13, 14], because these hydrogels also have nearly linear elasticity. However the native extracellular matrix, as well as the cytoskeleton in living materials is predominantly formed by relatively stiff fibrous polymer networks that exhibit a rich nonlinear viscoelastic response, with shear moduli that change by orders of magnitude over modest strains, and in some cases with frequency dependent changes in both the shear storage and loss (or elastic and viscous) moduli [15, 16]. Such nonlinear fibrous networks, composed for example of fibrin or collagen, can also be integrated with ferromagnetic particles to allow the elastic modulus of the composite to change by orders of magnitude very rapidly and reversibly by application of magnetic fields that can easily be generated in a laboratory setting [17, 18].

The importance of substrate stiffness and a schematic image of the methods used to study cell response to environmental viscoelasticity is illustrated in Fig. 11.1. Typically, the rigid glass or plastic substrate traditionally used for cell biology in the laboratory is covered by a thin elastomeric or hydrogel material on the surface of which specific adhesion proteins, typically extracellular matrix (ECM) proteins, are covalently attached [19, 20]. The elastic modulus of the deformable elastomer or hydrogel can be varied by altering polymer density, crosslinker concentration, and other features to vary the elastic modulus from less than 100 Pa to kilopascal or megapascal stiffnesses that span the range of most soft tissues, from brain to muscle to cartilage. An illustration of the importance of substrate mechanics is shown by the morphology of cardiac myocytes that are removed from the three-dimensional cardiac tissue and then placed on artificial surfaces with different stiffnesses [21]. Under chemically identical culture conditions, the morphology of these cells can vary from small and round to highly spread and polygonal at the two extremes of stiffness, but only on intermediate stiffnesses of 5–10 kPa, that mimic the stiffness of the native cardiac tissue, the cells acquire the elongated sarcomere-containing structures that allow them to rhythmically contract. In these and similar experiments, substrates of a constant stiffness are used for cell culture, but in living organisms

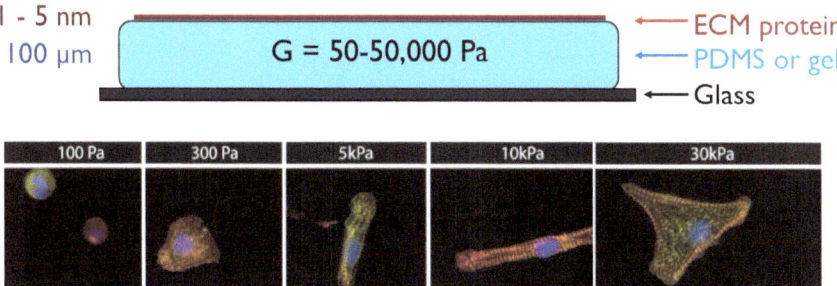

Fig. 11.1 Top: Diagram of the use of soft substrates with controlled shear modulus (G) for cell culture. bottom: Effect of substrate stiffness on the morphology of cardiac myocytes. Adapted with permission from Chopra A, et al. J Biomech. 2012;45:824-31. Copyright 2021 Elsevier, Inc.

tissue or extracellular matrix stiffness can change due to chemical remodeling or imposition of mechanical stresses. To achieve large, rapid, and reversible changes in substrate stiffness without chemically altering the substrate structure, magnetoelastic materials formed by embedding ferromagnetic particles into elastomers or hydrogels have recently been adopted for a variety of cell biological experiments.

Usually, these magnetoelastic substrates, like the example shown in Fig. 11.1, provide a surface on which cells can grow, and the elastic modulus of the substrate is altered by imposition of a magnetic field. A more recent advance has been to employ three-dimensional fibrous networks formed by the same protein filaments that form the extracellular matrices of many soft tissues, such as collagen or fibrin, and add ferromagnetic particles entrapped in the network meshes [17]. This allows cells to be cultured in a three-dimensional matrix that more closely mimics the setting of most cells in the body. Provided that the volume fraction of magnetic particles is sufficiently low, the particles themselves do not alter the structure or the rheology of the fibrous networks and provide the opportunity to change the effective stiffness that cells encounter in a three-dimensional network when a magnetic field is applied.

11.2   Rheological Properties of Dynamically Stiffening Soft Magnetoelastic Composites

In this study we summarize representative effects of applying uniform magnetic fields to elastomers and hydrogels containing ferromagnetic particles, with an emphasis on the strain dependence of the field-induced stiffening and a comparison of the differences between linear and nonlinear elastic networks. Analysis of the effect of magnetic fields on material stiffening shows that under the magnitudes of field strength and volume fraction of particles used in most such materials suitable for cell biology, the particles are largely immobile and trapped within the surrounding matrix, with the result that the stiffening effect is related to the generation of an array of magnetic dipoles within the network rather than application of local stress to the network. We illustrate the utility of magnetic stiffening in soft fibrous networks formed by collagen and fibrin, within which cells are embedded in a three-dimensional matrix. The rapid and reversible change in stiffness generated as the field is applied, without imposition of a local force on the cell, enables studies of both acute and chronic responses of cells to substrate stiffening. The most rapid response of the cell to a stiffened environment occurs within seconds and appears to involve activation of ion channels, that later lead to cell remodeling and changes in cell fate.

11.2.1   Magnitude of Magnetic Stiffening of Polydimethylsiloxane Containing Carbonyl Iron Particles

The magnitude of the stiffening in an elastomer substrate caused by a uniform magnetic field is seen in the examples shown in Fig. 11.2. In this study polydimethyl-siloxane (PDMS) elastomers with a shear modulus of approximately 5 kPa were formed with 10% weight fraction of randomly distributed carbonyl iron spheres with a diameter of approximately 3 microns (Sigma-Aldrich C3518). As seen in Fig. 11.2a the shear modulus increases from its initial value of 5 kPa to approximately 20 kPa in the presence of the 400 mTesla magnetic field. A theoretical model that computes the additional resistance to shear deformation provided by a random array of magnetic dipoles that mimic those that would be formed by the carbonyl iron particles predicts that the shear modulus should rise with a square of the magnetic field magnitude [18]. The fit of this theory to the experimental data shows very close agreement, from which the magnetic susceptibility of the particles can be computed. This theory also predicts that along with the resistance to shear deformation, increasing magnetic fields will generate a normal force within the material, and Fig. 11.2b shows that this normal force also rises with the square of the field strength, as predicted by the theoretical model (blue arrows). In these experiments the elastomer was placed between two rigid plates within a rheometer, but in settings in which a magnetic field is generated by a permanent magnet placed beneath it in a cell culture dish, the resulting normal force can lead to wrinkling of the upper surface of the elastomer to which the cells adhere [11]. The magnitude of this wrinkling effect, which could perturb cell adhesion to the surface, depends on the relative magnitudes of the normal force and the shear modulus of the elastomer. Direct measurements of elastomers suitable for cell culture show that the surface roughness caused by high fields is on the order of 10–100 s of nm [11], but does not seem to perturb the cell morphology.

Both theory and experiment show that the contribution of the ferromagnetic particle array in the magnetic field to the shear modulus is additive above the modulus of the elastomer in the absence of a field, or in the absence of ferromagnetic particles. The relative contributions from the magnetic particles and the underlying elastomer

Fig. 11.2 Effect of magnetic field on the shear storage modulus (**a**) and normal stress (**b**) when a magnetic field is applied to crosslinked polydimethylsiloxane containing 10% by weight carbonyl iron beads

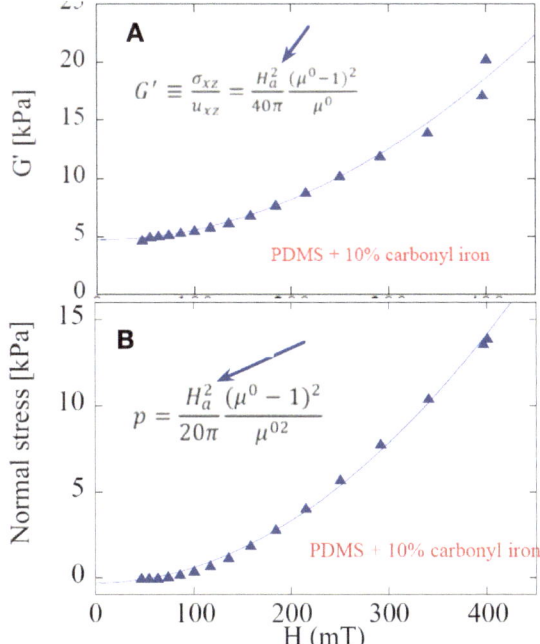

depend on the magnitude of the shear deformation. Figure 11.3 shows that an elastomeric material like PDMS, which exhibits nearly perfectly linear elastic response up to strains of at least 50%, becomes significantly strain softened after its stiffening by the magnetic field. In the presence of the particles but no magnetic field the shear modulus is nearly constant over the entire range of shear strains. After stiffening by the magnetic field, however, the shear modulus decreases significantly as shear strain magnitude is increased up to 100%. This softening is not the result of plastic deformation or damage to the network, because decreasing the shear strain magnitude to low values immediately leads to a higher value of measured shear modulus.

11.2.2   Theoretical Model for Stiffening of a Linearly Elastic Materials Containing Ferromagnetic Particles

The experimental results on magnetorheological properties of ferrogels can be rationalized by considering the dependence of their magnetic permeability tensor μ_{ik} on the deformation u [22]:

$$\mu_{ij} = \mu_0 \delta_{ij} + a_1 u_{ij} + a_2 u_{kk} \delta_{ij}. \tag{11.1}$$

Fig. 11.3 Shear modulus of magneto-elastomeric PDMS decreases with increasing strain in the presence of magnetic fields

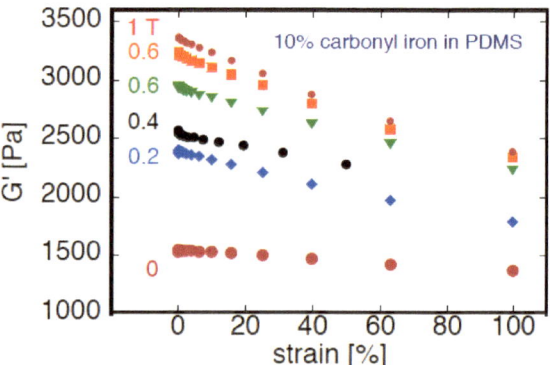

The magnetostriction coefficients a_1, a_2 are calculated in [23] using a self-consistent mean field approach (the magnetic susceptibility dependence on the particle concentration is taken into account in a_2) and are given as

$$a_1 = -\frac{2}{5}(\mu_0 - 1)^2; a_2 = -\frac{1}{5}(\mu_0 - 1)^2 \tag{11.2}$$

The coefficient a_2 plays a role only in compressible ferromagnetic media and will not be further considered.

Equations (11.1) and (11.2) allow calculation of the shear modulus dependence on the magnetic field. The volume density of magnetic torque determines the antisymmetric stress.

$$\sigma_{ij}^a = \tfrac{1}{2} e_{ijk} \left[\vec{M} \times \vec{H_0} \right]_k \quad [24].$$

In the magnetic field $\vec{H_0} = (0, 0, H_0)$ at a shear deformation $u_{xz} = \tfrac{1}{2}\frac{\partial u_x}{\partial z}$ a transversal component of the magnetization, M_x, arises:

$$M_x = \frac{a_1 H_0}{2\mu_0} \frac{\partial u_x}{\partial z}.$$

As a result, additional shear stress in the gel appears

$$\sigma_{xz}^a = \frac{a_1 H_0^2}{16\pi \mu_0} \frac{\partial u_x}{\partial z}$$

Thus, the effective shear modulus of the gel increases by the magnitude

$$\Delta G' = \frac{(\mu_0 - 1)^2 H_0^2}{40\pi \mu_0}$$

Another effect which is possible to measure by rheometry is the normal force acting on the plates holding the sample under the action of the normal field. The general expression for the magnetic energy of the body $-\frac{1}{2}\int \vec{M}\cdot \vec{H_0}dV$ [22] in the case of the gel layer with area S and thickness h gives

$$E = -\frac{1}{8\pi}H_0^2 Sh \frac{\mu_{zz}-1}{\mu_{zz}}$$

Its variation from the isotropic case at deformation $u_{zz} = \frac{\xi}{h}$ (ξ is the displacement of the upper plate of the rheometer) gives

$$\delta E = -\frac{a_1 H_0^2}{8\pi\,\mu_0^2}S\xi$$

Therefore, the force per unit area of the plate F is

$$F = -\frac{(\mu_0 - 1)^2 H_0^2}{20\pi\,\mu_0^2}$$

It may be noted that accounting for the elastic energy of an incompressible gel with volume V, $V\frac{3Gu_{zz}^2}{2}$ for shrinking deformation of the sample in the form of a disk of large radius we obtain

$$\left(M_0 = \frac{(\mu_0 - 1)H_0}{4\pi\,\mu_0} \right)$$

$$u_{zz} = -\frac{4\pi M_0^2}{15G},$$

This expression coincides with that derived in [25] for the description of the Procrustes effect. The quadratic dependencies of shear modulus increase and the normal force on the applied field correspond well with the experimental data.

11.2.3   Magnetoelastic Materials Formed by Fibrous Biopolymer Networks

Magnetoelastic materials formed by incorporation of ferromagnetic particles into polymeric elastomers have now been extensively used to study the effects of stiffness changes on induction of differentiation pathways in precursor cells, phenotypic changes in muscle cells, and other applications [6, 7, 9, 18, 26, 27]. Some limitations of solid elastomers are that cells can only be cultured on their surfaces, and the shear modulus of the elastomer before application of the field is generally above a few kPa, and stiffer than some of the softest tissues such as embryos, bone marrow, brain, or fat.

To circumvent the limitations of magnetoelastic elastomers, similar materials have also been formed by adding ferromagnetic particles to hydrogels. Some of the first examples were hydrogels formed by polyacrylamide or carrageenan, with initial elastic moduli below a kilopascal [6, 28]. Since the stiffening effect of the magnetic field on the ferromagnetic particles is additive to the initial elastic modulus of the host material, the fractional change produced by the same volume fraction of particles is much greater when the initial elastic modulus is low. Additionally, a change in elastic modulus that is adequate to alter cell phenotype can be produced by a lower volume fraction of ferromagnetic particles when the initial substrate stiffness is low.

To take a step closer to the native extracellular matrix environment of cells in three-dimensional cultures, magnetoelastic materials have also recently been made using the native biopolymer networks formed by fibrin or collagen that constitute the material into which cells infiltrate during wound healing or that surround the cell in homeostasis [17, 18]. In addition to providing a more native environment for the cell, the large mesh size and biocompatibility of fibrin and collagen enable cell culture in three-dimensional environments. An example of the formation of an optically translucent magnetoelastic material from a biopolymer is shown in Fig. 11.4. Carbonyl iron particles are suspended within culture medium prior to mixing with a solution of fibrinogen, the protein that polymerizes to form a blood clot after its activation by thrombin. Before fibrinogen is activated by the thrombin, the mixture can be poured into a mold or microfluidic chamber. Figure 11.4a shows a cylindrical fibrin gel with 0.5 percent carbonyl iron particles by weight. The sample is grey but partly transparent. The resulting fiber network entraps the carbonyl iron particles within it, as shown by the scanning electron micrograph in Fig. 11.4b.

Figure 11.4c shows that the shear modulus of a fibrin gel can be increased greatly by relatively modest magnetic fields. A fibrin gel with 10% carbonyl iron particles that has an initial shear modulus of 200 Pa is stiffened by a factor of 40 to over 8 kPa for the same magnitude of field that increased the stiffness of the initially stiffer PDMS elastomer by only a factor of 4 (Fig. 11.2). In addition, the nonlinear rheology of the semiflexible fibrin gel is also evident in the response of the magnetically stiffened fibrin gel to increasing shear strain magnitudes. Figure 11.4d shows that the initial decrease in shear modulus of the magnetically stiffened fibrin/bead composite switches to shear strain stiffening, at strains above 10%, consistent with the strong increase in shear modulus of the fiber network. Several aspects of the elastic response of magnetically stiffened fibrin gels are not yet explained by theoretical models. For example, the rise in shear modulus of the fibrin/bead composite does not follow a quadratic relation to the magnetic field strength, as seen for the magnetoelastic PDMS elastomer (Fig. 11.2). Similarly, the strain dependence of the magnetoelastic effect is also not evident from current theoretical models.

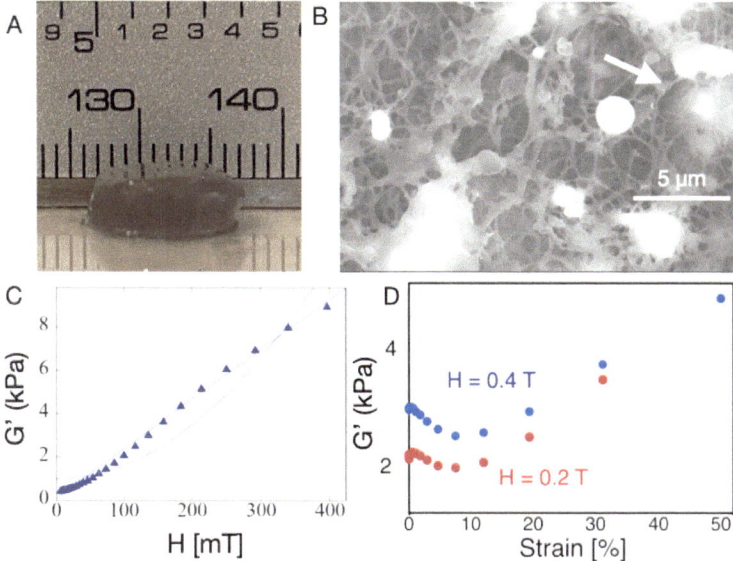

Fig. 11.4 a Photograph of a 10 mg/ml fibrin gel with 0.5% carbonyl iron beads. Scale: 1 mm between lines. **b** scanning electron micrograph of fibrin gel with ferromagnetic particle (arrow) at right. Scale bar: 5 μm. **c** effect of magnetic field on shear modulus of magnetoelastic fibrin gel. D. Strain dependence of magnetoelastic fibrin in presence of magnetic fields

11.3   Effects of Magnetoelastic Substrate Stiffening on Live Cells in 3D

The open fibrous meshwork of soft fibrin or collagen gels combined with the use of low volume fractions of carbonyl iron particles that permit imaging by light microscopy within the 3D network/cell composite creates new opportunities to study mechanobiology over a range of time scales in 3D environments that are close to the physiological setting. A schematic diagram of the method is shown in Fig. 11.5. The example shown here is for fibrin gels, but the same method can be used by adding cold acidic collagen in place of fibrinogen and initiating its polymerization by neutralization and warming to 37 °C as the cells are added. The utility of this system is demonstrated by effect of magnetic stiffening on the rapid change in intracellular Ca^{2+} flux when stiffness is suddenly changed.

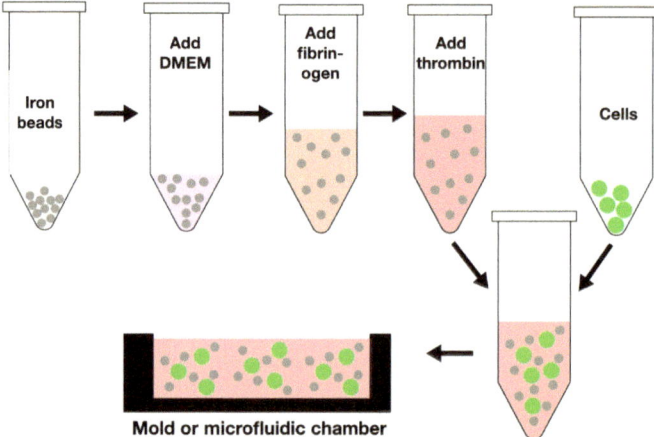

Fig. 11.5 3-D magnetoelastic cell culture system. Carbonyl iron particles are suspended in cell culture medium such as DMEM and then mixed with fibrinogen. The enzyme thrombin is then added at the same time that cells are added, and the polymerizing fibrin network can be poured into molds or microfluidic chambers before the fibrin gels

11.3.1 Magnetic Stiffening of Magnetoelastic Fibrous Networks Occurs in the Absence of Network Deformation by the Field

Figure 11.6a shows that fluorescently labeled live cells can be clearly imaged within a magnetoelastic collagen gel, and displacements of the beads can be used to measure a strain field caused by cells contracting the matrix. Figure 11.6b shows that the shear modulus of the magnetoelastic collagen gel containing only 0.5% carbonyl iron increases from ~ 0.5 to 1.6 kPa when a 400 mT field is applied and returns to its baseline level with the field is removed. The carbonyl iron microparticles can also be visualized within the hydrogel using fluorescent labeling to assess whether application of the magnetic fields causes displacement independent of cell contractility. As suggested by the theoretical model for stiffening in linear elastic materials, the contribution of the microparticles to the shear modulus of the fibrous network is predicated on the dipoles being entrapped in the network. Figure 11.6c demonstrates that application of a 250 mT field induces displacement of the particles prior to hydrogel polymerization, but there is minimal particle displacement once the collagen has polymerized. This result verifies that the field does not displace the particles once they are entrapped in the biopolymer network. In stiffer materials like PDMS, there is less potential for microparticles to move in the presence of a field. However, in biological materials like collagen and fibrin, the shear moduli may be too low to constrain the particles, mitigating the effect of the microparticles on network stiffness. These results indicate that at least in the case of 2 mg/mL collagen (storage modulus ~ 30–50 Pa), a magnetic field that is large enough to substantially increase hydrogel

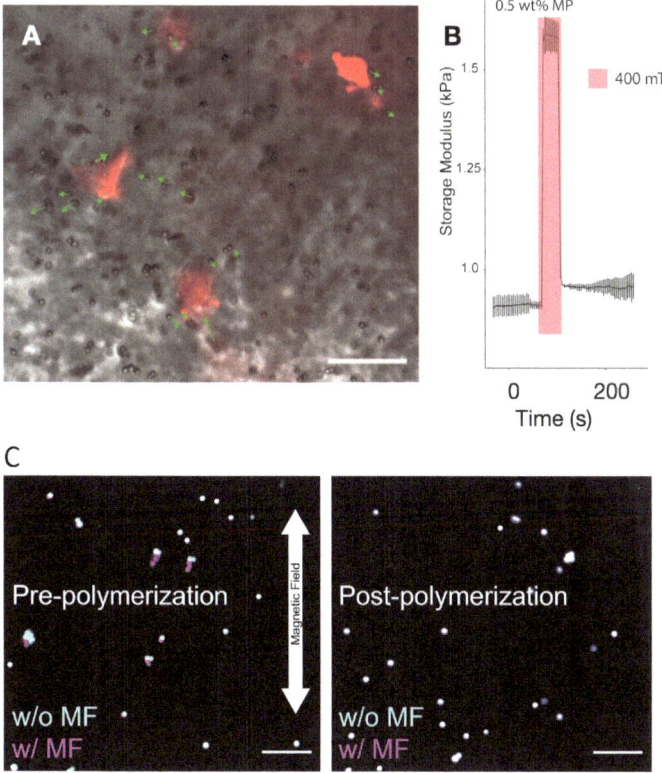

Fig. 11.6 a ifeACT-transfected hCASMC in hydrogels consisting of 5 mg/mL collagen, 1 mg/mL HA, and 0.5 wt% carbonyl iron particles. hCASMC are labeled with the live molecular probe, LifeACT (red), and green denotes the local strain caused by cell contractility. **b** change in shear modulus of hydrogel as uniform 400 mT field is applied and removed. **c** fluorescently labeled microparticles in a 2 mg/mL collagen hydrogel before and after polymerization in the presence of a 250 mT field. Scale = 50 μm. Adapted with permission from K.A. Tran, et al., ACS Appl. Mater. Interfaces **13**, 20,947–2094759 (2021). Copyright 2021 American Chemical Society

stiffness does not cause microparticle displacement. However, future studies using softer hydrogels or larger fields should verify that the microparticles are not displaced by application of the field.

11.3.2   Rapid Response of Cells to Sudden Stiffness Changes Involves Calcium Ion Fluxes

The ability to image cells within biopolymers containing carbonyl iron microparticles and the rapid and reversible changes to network mechanical properties made possible by the application of a magnetic field provide a unique glimpse into the dynamic

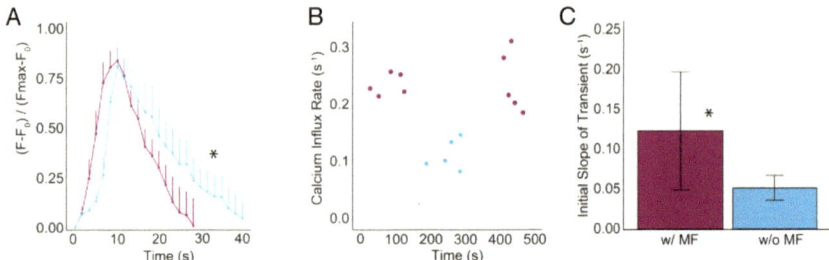

Fig. 11.7 Effects of dynamically altered hydrogel mechanics on cellular calcium transients. **a** Average calcium transients in carbonyl iron–seeded collagen hydrogels with field on (red) and off (blue). **b** calcium influx rate (slope of calcium transients) as magnetic field is turned on and off every 150 s. **c** average initial slope of calcium transients in matrices with or without stiffening by magnetic field. *$P < 0.01$. (n = 5 cells per condition, 10 transients for MF on and 5 transients for MF off per condition). Adapted with permission from K.A. Tran, et al., ACS Appl. Mater. Interfaces **13**, 20947–20959 (2021). Copyright 2021 American Chemical Society

cell response to extracellular matrix stiffness. As Fig. 11.6b indicates, the hydrogel storage modulus is increased to a steady value within milliseconds of the application of the magnetic field. Previous studies have used this rapid change to characterize how quickly cells respond to a shift in extracellular matrix mechanics [18]. Figure 11.7a shows averaged calcium transients from cells treated with Fluo-4, a calcium sensitive fluorescent dye, in the presence of an intermittent magnetic field. The transient is significantly different during the time when the field is applied (red) compared to when there is no field (blue): both in the overall rate as well as the initial slope of the transient (Fig. 11.7b, c). This result indicates the near instantaneous response of the cells to the field, as well as the reversibility of this response. There does not appear to be any inertia in the calcium flux in the cells: once the field is removed and the hydrogel stiffness decreases, the transients return to baseline levels.

11.4   Conclusion

These results validate the ability of magnetically active biopolymer hydrogels to interrogate the dynamics of cell mechanics. The effect of the magnetic field on calcium handling demonstrates that cells respond rapidly to changes in hydrogel stiffness, but a host of questions remain about whether the acute response is limited to cytoskeletal-mediated changes or whether gene transcription or translation is also affected in the following seconds and minutes. Future work can also include studies to determine the effect of the magnetic field on the viscoelastic properties of biopolymers like collagen and fibrin. In contrast to linear elastic materials like PDMS and polyacrylamide, collagen and fibrin have substantial viscous dissipation that is also affected by the application of a magnetic field. Using magnetic fields to tune the viscoelastic properties would provide new avenues to understand cell response to changes in their physical surroundings.

Acknowledgements This work was supported by the US National Science Foundation Center for Engineering Mechanobiology CMMI-1548571 (KP, PAJ) and NSF #2223318 (PAG, KAT), the US National Institutes of Health EB017753 (PAJ, KP) and GM136259 (PAJ), and the Scientific Council of Latvia grant No.lzp-2020/1-0149 (AC).

References

1. P.A. Janmey, D.A. Fletcher, C.A. Reinhart-King, Physiol. Rev. **100**, 695 (2020)
2. K.H. Vining, A. Stafford, D.J. Mooney, Biomaterials **188**, 187 (2019)
3. J.A. Burdick, W.L. Murphy, Nat. Commun. **3**, 1269 (2012)
4. R.S. Stowers, S.C. Allen, L.J. Suggs, Proc Natl Acad Sci U S A **112**, 1953 (2015)
5. A.M. Rosales, C.B. Rodell, M.H. Chen, M.G. Morrow, K.S. Anseth, J.A. Burdick, Bioconjug. Chem. **29**, 905 (2018)
6. A.A. Abdeen, J. Lee, N.A. Bharadwaj, R.H. Ewoldt, K.A. Kilian, Adv Healthc Mater **5**, 2536 (2016)
7. W. Chen, Y. Zhang, J. Kumari, H. Engelkamp, P.H.J. Kouwer, Nano Lett. **21**, 6740 (2021)
8. C. Gila-Vilchez, J.D.G. Duran, F. Gonzalez-Caballero, A. Zubarev, M.T. Lopez-Lopez, Smart Mater. Struct. **28**, 035018 (2019)
9. M. Mayer et al., PLoS ONE **8**, e76196 (2013)
10. K.M. Meyers, K.G. Ong, Sustainability **13**, 13655 (2021)
11. A.T. Clark, A. Bennett, E. Kraus, K. Pogoda, P. Cēbers, P. Janmey, K.T. Turner, E.A. Corbin, X. Cheng, Multifunc. Mat. **4**, 035001 (2021)
12. D. Garcia-Gonzalez, M.A. Moreno, L. Valencia, A. Arias, D. Velasco, Comp. Part B-Eng. **215**, 108796 (2021)
13. T. Mitsumata, A. Honda, H. Kanazawa, M. Kawai, J. Phys. Chem. B **116**, 12341 (2012)
14. C. Xu, B. Li, X. Wang, Int J Mol Sci **22** (2021)
15. A.S.G. van Oosten, X. Chen, L. Chin, K. Cruz, A.E. Patteson, K. Pogoda, V.B. Shenoy, P.A. Janmey, Nature **573**, 96 (2019)
16. A. Sharma, A.J. Licup, R. Rens, M. Vahabi, K.A. Jansen, G.H. Koenderink, F.C. MacKintosh, Phys. Rev. E **94**, 042407 (2016)
17. G. Chaudhary, N.A. Bharadwaj, P.V. Braun, R.H. Ewoldt, ACS Macro Lett. **9**, 1632 (2020)
18. K.A. Tran, E. Kraus, A.T. Clark, A. Bennett, K. Pogoda, X. Cheng, A. Cēbers, P.A. Janmey, P.A. Galie, ACS Appl. Mater. Interfaces **13**, 20947 (2021)
19. C.E. Kandow, P.C. Georges, P.A. Janmey, K.A. Beningo, Method Cell Biol. **83**, 29 (2007)
20. K. Pogoda, E. E. Charrier, P. A. Janmey, Bio-Protocol **11** (2021)
21. A. Chopra et al., J. Biomech. **45**, 824 (2012)
22. L.D. Landau, E.M. Lifshitz, *Electrodynamics of Continuous Media* (Pergamon Press, 1984)
23. Y.M. Shkel, D.J. Klingenberg, J. Appl. Phys. **83**, 7834 (1998)
24. E. Blum, A. Cebers, M.M. Maiorov, *Magnetic Liquids* (W de Gruyter, Berlin, 1997)
25. K. Morozov, M. Shliomis, H. Yamaguchi, Phys. Rev. E **79** (2009)
26. E.G. Popa, V.E. Santo, M.T. Rodrigues, M.E. Gomes, Polymers **8**, 28 (2016)
27. M.S. Ting, J. Travas-Sejdic, J. Malmstrom, J. Mat. Chem. B **9**, 7578 (2021)
28. J. Ikeda, D. Takahashi, M. Watanabe, M. Kawai, T. Mitsumata, Gels **5**, 39 (2019)

Chapter 12
Magnetic Tape Head Tweezers for Novel Protein Nanomechanics Applications

Rafael Tapia-Rojo

Over the past few decades, magnetic tape heads have been perfected to allow for the application of strong magnetic fields that can be modulated at very high frequencies, meeting the technological demands of high-density magnetic recording. Hence, exploring their implementation in modern magnetic tweezers force spectroscopy seems like a natural approach. Here, recent developments in magnetic tweezers instrumentation related to the use of magnetic tape heads will be reviewed. Classic magnetic tweezers technology typically employs a pair of permanent magnets to apply pulling forces on tethered molecules, and force changes require to physically displace the magnets, a slow and often inaccurate process. By controlling force through the electric current supplied to the tape head, novel magnetic tweezers designs overcome this limitation, enabling swift force changes and opening new avenues to explore protein nanomechanics under rapidly changing forces. To illustrate the potential of this instrumental approach, two practical examples, respectively studying protein folding over short timescales and under complex force signals, will be discussed.

12.1 Introduction

Mechanical forces are a key player across all biological scales. At the macroscopic scale, our bodies are entirely accustomed to responding to mechanical forces—walking, lifting a weight, or playing sports. Similarly, mechanical forces play a fundamental role at smaller biological scales, including tissues, cells, and even single molecules [1]. Common physiological processes such as muscle contraction, tissue integrity, or cell motility are finely regulated by mechanical forces. At the molecular

R. Tapia-Rojo (✉)
Physics Department, King's College London, London, UK
e-mail: rafael.rojo@kcl.ac.uk

B. Doudin et al. (eds.), *Magnetic Microhydrodynamics*, Topics in Applied Physics 120,
https://doi.org/10.1007/978-3-031-58376-6_12

level, many of these processes are underpinned by force-bearing proteins, with the ability to detect and respond to mechanical cues to trigger signaling pathways that eventually result in the modulation of mechanosensitive gene-expression programs [1, 2]. In this sense, understanding the molecular underpinnings of such force-sensing processes requires measuring the nanomechanical response of the key protein players. While classical biochemical assays have provided a wealth of knowledge about protein function, unfortunately, they are of little help here simply due to their inability to apply forces to a collection of proteins in the bulk.

In this context, single-molecule force spectroscopy techniques enable us to subject individual proteins to pN-level forces and measure their force-dependent conformational dynamics, which underpin, in many cases, their function [3, 4]. In a nutshell, force spectroscopy techniques—namely atomic force microscopy (AFM), optical tweezers, and magnetic tweezers—are based on the same concept: anchoring an individual molecule between its termini to apply a stretching force while monitoring its dynamics in terms of end-to-end length changes. For the study of protein nanomechanics, AFM and optical tweezers have traditionally been the techniques of choice [5], while magnetic tweezers have been mostly devoted to studying nucleic acids [6]. This likely owes to the low temporal and spatial resolution of the early magnetic tweezers designs, sufficient to capture the dynamics of the very stiff and long DNA molecules, in contrast to the more subtle nanomechanical behavior of most proteins. However, recent instrumental advances and the development of novel chemical strategies for protein anchoring are now demonstrating the expediency of magnetic tweezers for measuring protein nanomechanics [7–10].

Magnetic tweezers offer some natural advantages for studying protein dynamics under force. First, it is an intrinsically very stable technique, which has recently achieved very long measurements of protein dynamics over several hours or even days, hence allowing us to monitor single proteins over physiologically relevant timescales [7, 8, 10, 11]. Additionally, and in contrast to AFM and optical tweezers that generate force by displacing a force-probe, magnetic tweezers offer intrinsic force clamp conditions due to the very soft trap created by the magnetic field (typically $\sim 10^{-6}$ pN/nm). This affords direct control of the intrinsic variable (force) while the extensive variable (molecular extension) is measured, hence providing the natural statistical ensemble to measure protein dynamics in equilibrium. However, despite these advantages, being able to apply calibrated forces and, more specifically, to accurately and quickly change them has been a classical limitation in magnetic tweezers instrumentation. In its most typical configuration, magnetic tweezers use a pair of permanent magnets placed on top of the fluid chamber containing the proteins tethered to superparamagnetic beads. While this simple strategy excels at applying constant forces (provided that the magnets are accurately positioned without any vertical drift), changing the force involves a physical displacement of the magnets, an intrinsically slow process that greatly limits how fast force can be modulated. This technical caveat prevents, for example, capturing early protein folding events that can occur shortly after fast force quenches or studying protein dynamics under complex force protocols, limiting the scientific scope of this technique.

This chapter will review recent advances in magnetic tweezers instrumentation focused on implementing a magnetic tape head as the force-generating apparatus. This technical development provides accurate control of the pulling force, which can now be changed very rapidly (~10 kHz), granting access to previously inaccessible force protocols for studying protein nanomechanics. The chapter starts with a brief instrumental revision, stressing the calibration problem and how this is addressed when using the magnetic tape head. Finally, two practical applications that take advantage of the tape head's features are presented, both highlighting the potential of this technique to undercover new properties in the response of individual proteins to force.

12.2 Magnetic Tape Head Tweezers—Instrumental Description and Calibration

12.2.1 Implementation of a Tape Head in a Single-Molecule Magnetic Tweezers Design

In magnetic tweezers, magnetic fields are used to apply forces of a few piconewtons (pN) to individual molecules tethered to micron-sized superparamagnetic beads. By tracking the vertical position of the bead (typically recording its interference or diffraction ring patterns), the molecular extension is measured, which allows for studying single-molecule dynamics under constant-force conditions [6, 8].

In most instrumental designs, the magnetic field is applied using a pair of permanent magnets placed on top of the fluid chamber, which directly exposes the magnetic beads to a constant force. By vertically displacing the magnets using a motor or a voice coil, the force is changed. The accessible range of forces depends on the magnets' (geometry, material, etc.) and the bead's properties (diameter, maximum magnetization, composition, etc.) While forces of several pN can be reached with some magnets-bead combinations [8, 12], magnetic tweezers are naturally suited to accurately manipulate low forces (<15 pN), in contrast to other techniques like AFM. This makes magnetic tweezers particularly appropriate for studying nucleic acids (typically very stiff molecules, which implies that most relevant conformational changes occur at very low forces) or proteins with low mechanical stability.

Recently, we introduced a new instrumental magnetic tweezers design, which implements a magnetic tape head as the force-generating apparatus [7, 10]. Magnetic tape heads have been developed and employed for decades in the perhaps now obsolete field of magnetic head recording [13]. Magnetic recording requires the application of intense magnetic field pulses at very high rates, essential for achieving high-density recording. This technological necessity pushed the industry to develop tape head devices with the ability to change the magnetic field very rapidly and with minimal thermal dissipation. In our context, this capability is particularly convenient as it overcomes one of the intrinsic limitations of magnetic tweezers: changing the force at high rates. Since most magnetic tweezers instruments use permanent magnets, force changes require physically displacing them, a generally slow process, taking up to ~ 100 ms in the best-case scenario. Due to this limitation, fast molecular events can be lost during the force change, where the force is ill-defined. Additionally, this limitation has restricted the set of applicable force protocols to either constant force pulses or slowly changing force ramps, limiting our understanding of how proteins can respond to more complex force signals.

A magnetic tape head is simply an electromagnet consisting of a toroidal core of magnetic material with large permeability, split by a narrow gap (generally containing a diamagnetic material) over which a strong magnetic is generated when applying electric current to the coil tightly wrapped around the tape head's core [13]. In our instrumental design, we implemented a commercial magnetic tape head (Brush Industries, 9022836), which we selected to achieve forces comparable to those reached with the permanent magnets approach [8]. Among other commercially available tape heads, this specific model has a high maximum gap field (0.5 T) and strong gradient owing to its narrow gap (25 μm) that makes it suitable for our application, providing a working force range between 0 and 44 pN when applying electric currents between 0 and 1000 mA, above which the tape head saturates. To control the magnetic field at high rates, we connect the tape head to a custom-designed current-clamp PID circuit that maintains under feedback a high-precision 2 Ω resistor connected in series with the tape head (Fig. 12.1a). Our interest is being able to manipulate the magnetic field (force) only by controlling the electric current; hence, it is critical to position the tape head accurately at a fixed position over the fluid chamber. In particular, the magnetic field changes in the vertical coordinate over a length scale defined by the head's gap width (25 μm in our case), which requires positioning the tape head with micron accuracy. To this aim, we designed a mounting piece manufactured using high-precision CNC [10] (Fig. 12.1b). When mounted in this piece, the tape head is positioned 450 μm away from the bottom surface (this is, 300 μm away from the magnetic beads when using standard 150 μm-thick bottom glasses to assemble the fluid chambers), which allows the accurate application of mechanical forces between 0 and ~ 44 pN only by controlling the electric current. By assembling the mounting piece on top of an inverted microscope, the vertical position of the magnetic beads can be tracked using a standard image-analysis algorithm [8], which allows for measuring the molecular extension of single proteins subjected to mechanical forces.

12.2.2 Anchoring Chemistry

Developing specific and stable chemical anchors is a crucial endeavor in single-molecule force spectroscopy. Over the past few years, there has been significant effort in designing new strategies for tethering protein constructs for magnetic tweezers [8, 9]. Among these, the HaloTag chemistry has been proven to be one of the most effective ones, as it allows for covalent and highly specific anchoring of individual

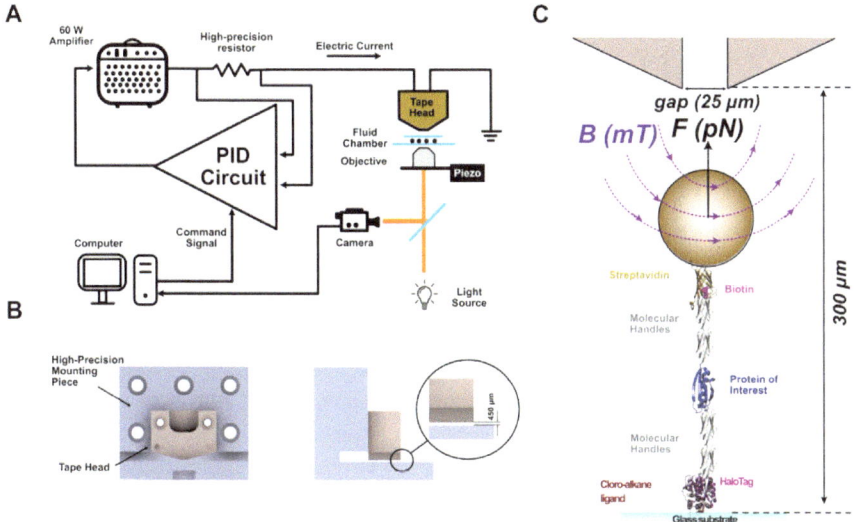

Fig. 12.1 Implementation of a magnetic tape head on a single-molecule magnetic tweezers instrument. a Schematics of the electronics and microscope design. The tape head is connected in series with a high-precision resistor, which is maintained under feedback with a current-clamp PID circuit powered by a 60 W amplifier. With this strategy, the electric current through the tape head is controlled, and, hence, the mechanical force. The tape head is positioned above the fluid chamber, placed on top of an inverted microscope, which allows for measuring the molecular extension. **b** The tape head is mounted on a high-precision piece fabricated with CNC, which allows positioning it 450 μm away from the bottom surface, hence, 300 μm from the magnetic beads, when using standard #1 bottom glasses. **c** Schematics of the molecular anchoring. The protein of interest is flanked by two stiff protein domains (typically Ig32, or Spy0128), which serve as molecular anchors. At the N-terminus, a HaloTag allows covalent anchoring of the construct to a glass substrate functionalized with the HaloTag Amine O4 ligand. At the C-terminus, a biotinylated AviTag closes the tether by interacting with streptavidin-coated superparamagnetic beads. The tape head is positioned 300 μm away from the bead/glass substrate, which allows the application of magnetic fields in the ~ mT range, which generate highly-controlled pN-level forces (diagram not at scale, the bead's diameter is ~ 2.8 μm, while the protein construct extension is ~ 20 nm)

proteins to glass coverslips that can be functionalized following a simple protocol [8, 14]. Tethering the protein construct to the superparamagnetic bead relies on the chemical coating of the commercially available beads. In the simplest and most used approach, the protein construct is capped with a biotinylated AviTag that permits binding to commercially available streptavidin-coated beads. The streptavidin–biotin interaction, albeit not covalent, is strong enough to resist forces of a few pN (< 65 pN) for extended times. However, if higher forces are required, it is possible to use double-covalent anchors by combining HaloTag chemistry with the SpyCatcher-SpyTag protein fusion system [15].

Another aspect to consider, particularly when studying protein monomers, is to develop suitable molecular handles to space the protein interest from the glass surface and bead to avoid spurious non-specific interactions. While DNA handles are often used in many applications, in our approach, we flank the protein of interest between two mechanically stiff protein domains (typically the titin Ig32 domain), which require very high forces to unfold, hence, not interfering with the dynamics of the protein of interest.

12.2.3 Calibration

An intrinsic challenge in any single-molecule force spectroscopy technique is accurately determining the applied force. In magnetic tweezers, this requires relating the applied magnetic field with the force acting on the bead to which the single protein is tethered. When using permanent magnets, while developing an analytical expression for the magnetic force is possible (see, for example, [16, 17] for discussions in this regard), an empirical exponential dependence is typically employed [8]. Although this approach lacks a rigorous physical basis, it works in practice, at least within the required force precision and range.

For a magnetic tape head, there is a general expression for the magnetic field as a function of the electric current and distance to the tape head's gap; the so-called Karlqvist approximation predicts the magnetic field as [11, 18, 19]:

$$B_x = \frac{B_g}{\pi} \left[\tan^{-1} \left(\frac{g/2 + x}{z} \right) + \tan^{-1} \left(\frac{g/2 - x}{z} \right) \right] \tag{12.1}$$

$$B_z = \frac{B_g}{2\pi} ln \frac{(g/2 + x)^2 + z^2}{(g/2 - x)^2 + z^2}, \tag{12.2}$$

where B is the magnetic field, x is the lateral coordinate and z is the vertical one (we assume symmetry in y), g is the width of the tape head's gap, and B_g is the gap field, which depends on the properties of the tape head:

$$B_g = \mu_0 \frac{NI}{g} \eta, \tag{12.3}$$

being N the number of wire turns around the head core, I the applied intensity, $\mu_0 = 4\pi \times 10^{-7}$ Tm/A the vacuum permeability, and η the field efficiency, defined by the specific geometry of the tape head. For a given magnetic field $\vec{B}$, the force acting on a superparamagnetic bead is:

$$\vec{F} = (\vec{m} \cdot \nabla) \vec{B} \tag{12.4}$$

Since, in practice, we cannot neglect an initial magnetization of the bead M_0, the total magnetization of the bead [20]:

$$\vec{M} = \vec{M_0} + \frac{\chi_b}{\rho} \frac{\vec{B}}{\mu_0}, \tag{12.5}$$

where ρ is the density of the bead and χ_b its initial susceptibility. Thus, the magnetic force acting on the bead:

$$\vec{F} = \rho V \left(\vec{M_0} \cdot \nabla \right) \vec{B} + \frac{V \chi_b}{\mu_0} \left(\vec{B} \cdot \nabla \right) \vec{B}, \tag{12.6}$$

being V the bead's volume. When working right under the tape head's gap, there is no lateral component of the field gradient, so the only force component will be in z (pulling force), which is the desired experimental situation. Here, we can write the pulling force as a function of the distance z and the electric current I [7]:

$$F(z, I) = AI^2 \tan^{-1} \left(\frac{g/2}{z} \right) \frac{1}{1 + \left(\frac{z}{g/2} \right)^2} + BI \frac{1}{1 + \left(\frac{z}{g/2} \right)^2}, \tag{12.7}$$

where A and B are coefficients that depend on the bead's and head's properties, namely:

$$A = 8V \chi_b \mu_0 \frac{N^2}{g^3} \frac{\eta^2}{\pi^2}$$

$$B = 4\rho \mu_0 \frac{N}{g^2} \frac{\eta}{\pi} M_{0x}$$

Therefore, provided we know g, we can leave A and B as free parameters to determine our force calibration curve (they are difficult to calculate as they depend on magnitudes such as η, which are hard to estimate). To determine A and B, we need some molecular quantity whose force dependence is well-described and easily measurable. In our case, we employ the force-dependent extension changes of a folding/unfolding protein (step sizes) as a molecular ruler to relate force with the magnetic field [8].

When a protein unfolds under force, it becomes an unstructured polypeptide that quickly equilibrates to an average extension ($<x>$) that depends on the pulling force (F) following standard polymer physics models, such as the freely-jointed chain (FJC) model [7, 8, 21]:

$$<x> (F) = \Delta L_C \left[coth \left(\frac{F l_K}{kT} \right) - \frac{kT}{F l_K} \right], \tag{12.8}$$

where $kT = 4.11$ pN nm is the thermal energy, ΔL_C the change in contour length (total extension of the unfolded protein minus the size of the folded structure), and l_K the Kuhn length (related to the protein stiffness, for an unfolded protein $l_K = 1.1$ nm).

Hence, $< x > (F)$ provides a good observable, which is simple to measure experimentally, and that, from its force dependence, allows calibrating by determining the parameters A and B. If we position the tape head at a fixed distance, in our case $z = 300$ μm, simply by measuring the average step sizes $< x >$ as a function of the applied electric current I, we can determine the calibrating parameters A and B.

We used a protein L octamer construct (eight identical repeats of the bacterial protein L protein arranged in tandem) as a molecular ruler [7] (Fig. 12.2a). Protein L is a standard protein folding model that exhibits a clear folding/unfolding signature over a broad range of forces, providing a robust molecular observable to determine the calibration parameters A and B. Based on previous studies, the elastic properties of protein L have been accurately described by the FJC model with $\Delta L_C = 16.3$ nm and $l_K = 1.1$ nm [22].

Figure 12.2b shows a typical recording of a protein L octamer using magnetic tweezers. First, we applied a high electric current pulse of 932 mA, which readily unfolds the eight domains, appearing as discrete ~ 14.5 nm steps that increase the protein extension. We then lowered the current (force) to 371 mA, and the extended polypeptide collapsed under force, to then show stochastic folding/unfolding dynamics of its domains, appearing as downward (red arrow, folding) and upward (blue arrow, unfolding) steps, with a change in extension of ~ 8.5 nm. With this procedure, we explored different values of I and measured the step sizes of protein L to relate them with the pulling force, given its well-known molecular properties. Figure 12.2c shows the step sizes of protein L as a function of the applied electric current I fitted to Eq. 12.9, which provides the calibration parameters: $A =$

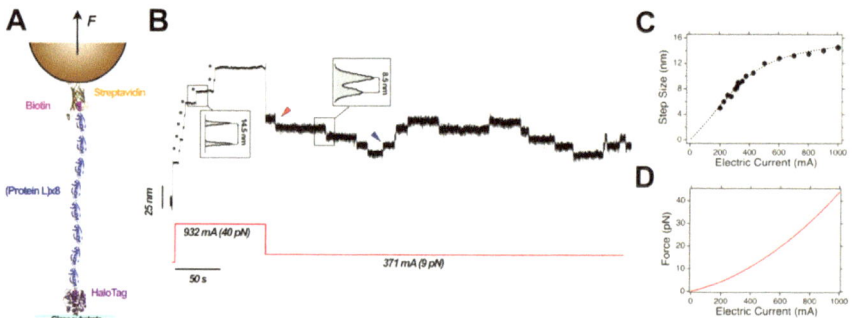

Fig. 12.2 Calibration Strategy for Single-Protein Magnetic Tweezers. a Schematics of a protein L octamer construct for magnetic tweezers. **b** Representative magnetic tweezers recording of a protein L octamer. A high current (force) pulse, unfolds the full polyprotein, showing eight discrete steps associated with the extension of each protein L domain (stars). A subsequent lower current (force) pulse reveals the reversible folding/unfolding dynamics as downwards (red arrow) and upwards (blue arrow) steps. The step sizes scale with force following the freely-jointed chain model, which allows us to associate electric current with the pulling force. **c** Step sizes of protein L as a function of the applied current. Error bars are the standard deviation. The dotted line is the fit to Eq. (12.8), which provides the calibration parameters A and B. **d** Force law as derived from **c**. Using the M270 beads and positioning the tape head at 300 μm from the surfaces enables the application of forces between 0 and 44 pN

$(2.952 \pm 0.853) \times 10^{-5}$ pN/mA^2 and $B = 0.016 \pm 0.005$ pN/mA. These parameters define our *current law* that allows us to calculate the pulling force as a function of the applied current (Fig. 12.2d), provided that the tape head is fixed at a position of z = 300 μm. A more general calibration, probing different tape head positions z was described in Ref. [7].

The values of A and B depend on the specific magnetic tape head model and the beads' properties (here, Dynabeads M270); therefore, the parameters discussed here are specific to our configuration. Using tape heads with stronger gap fields or bigger beads would allow for a broader range of forces. For instance, the commercially available M450 beads have also been proved in our experimental system, allowing us to reach forces up to ~ 240 pN [12].

12.3 Applications

Here, two practical examples highlighting the application of our tape head-based magnetic tweezers approach will be discussed.

12.3.1 Dissecting the Folding Pathway of a Single Protein on the Millisecond Timescale

The protein folding problem—this is, how an unfolded protein statistically samples its very large conformational space to find the unique native folded state—remains a key question in biophysics, which still attracts significant experimental and computational efforts [23, 24]. When folding, it is now generally accepted that the unfolded polypeptide will traverse a progressively narrower conformational space (in the entropic sense) that will eventually lead to the unique native folded state that minimizes the protein's free under the given folding conditions. This is known as the now-classic funnel vision of protein folding [22, 25]. In this context, the first stage in the folding pathway is assumed to be an entropy-driven hydrophobic collapse of the polypeptide, where all the non-polar side chains clump together away from the polar solvent. This still immature protein structure is generally known as the molten globule state [26]. From the molten globule state, the protein conformation eventually evolves to establish the specific and unique interactions that define the native structure—such as hydrogen bond networks, salt bridges, etc.

Although this vision of protein folding agrees with statistical mechanics and computational models and simulations, it has been challenging to demonstrate it from an experimental perspective. In biochemical bulk studies, the diversity of folding reactions in an ensemble of folding proteins is averaged out, typically rendering the classic two-state picture of protein folding. On the other hand, while force spectroscopy studies enable direct observation of the folding pathway of an individual protein, the collapse transition to the molten globule is typically very fast, being invisible to most experimental techniques. Previously, an on-pathway molten-globule state in the RNase H protein was captured using optical tweezers [27]; however, this mechanically labile state was strikingly long-lived (~seconds), allowing its direct visualization. Over the past few years, instrumental developments have allowed a higher-resolution inspection of folding proteins. The development of high-speed force-clamp AFM enabled the detection of an ensemble of mechanically weak states in folding ubiquitin polyproteins that preceded the native folded state [28]; however, the AFM has limitations in the low-force range, which prevents precisely ascertaining the folding forces and thus, to directly monitor the folding reaction.

The implementation of the magnetic tape head overcomes many of these limitations by now allowing us to carry out swift force changes in the microsecond timescale and thus to finely sample the folding pathway of an individual protein. Here, we used protein L as a classic protein folding model, widely reported to fold in an uncomplicated two-state fashion [22, 29]. In these experiments [7], we designed a force pulse protocol to allow a protein L octamer to fold at a very low force during a precisely controlled time Δt and then interrupted this reaction with a higher force pulse to interrogate the folding status of the protein. Figure 12.3a illustrates the pulse protocol: We started with an unfolding pulse at 40 pN, which readily unfolds all eight protein L domains, to then decrease the force to 10 pN, which sets the reference extension of the unfolded polyprotein at this force. We then lowered the force to 1 pN to allow folding during a time Δt (here 250 ms) and quickly interrupted the folding pathway with a second 10 pN pulse to probe the folding status of the protein L octamer.

During the probe pulse, we can clearly distinguish between two different kinds of collapsed states by their mechanical stability. Shortly after the force increases to 10 pN, we can observe very rapidly unfolding events on a milliseconds timescale, indicative of mechanically labile protein states (Fig. 12.3a, inset, red arrows); then, over a longer timescale, there is an additional unfolding event that occurs at the expected rate for folded protein L at 10 pN (Fig. 12.3a). We repeated this protocol on several protein L molecules and measured the unfolding times of all events observed during the 10 pN probe pulse. The unfolding times—here plotted with logarithmic binning to facilitate identification of the involved timescales—are distributed as a double exponential, with a short timescale of $\tau \sim 0.5$ s and a longer timescale of $\tau \sim 50$ s (Fig. 12.3b). This mixed population of structures with such different mechanical stabilities indicates the presence of two different protein structures that are attained over the brief 250 ms-long folding pulse, likely a mechanically labile molten-globule-like state and the mechanically stiff folded state of protein L.

To understand the nature of this state, we varied our folding time Δt to sample the folding pathway of protein L. Figure 12.3c shows the probability of finding a protein domain in the unfolded state (no step), molten-globule (fast unfolding), and native (slow unfolding), which dissects the time-evolution of the folding pathway of protein L. Over 10–25 ms, we found only unfolded domains, indicating that over these short times, protein L has simply no time to reach any mechanically stable state. As we increased the quench time to 100–500 ms, we observed a maximal population of

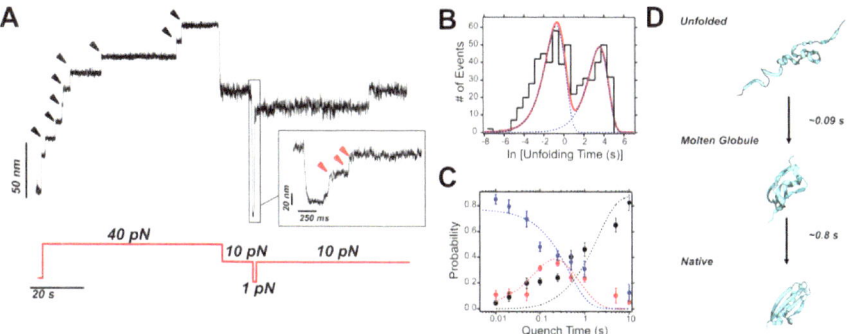

Fig. 12.3 Direct detection of ephemeral molten globule states in the folding pathway of protein L. a Magnetic tweezers recording illustrating the pulse protocol for detecting molten globule states. **b** Square root histogram showing the unfolding kinetics of protein L at 10 pN after a brief quench pulse at 10 pN. The histogram clearly separates two populations, a set of mechanically labile states that unfold over a timescale of ~ 0.5 s (first peak) and a mechanically stiff set unfolding over ~ 50 s (second peak). **c** Probability of reaching the molten globule (red), native state (black), or remaining in the unfolded state (blue) as a function of the quench time. The data is well described with a first-order three-state kinetic model that allows extracting the folding rates of protein L. **d** Schematic description of the folding pathway of protein L. Starting from the unfolded state, the protein rapidly transitions to a mechanically labile molten globule-like state over ~ 0.09 s, followed by an entropy-driven hydrophobic collapse. Over a slower timescale of ~ 0.8 s, the protein forms the enthalpic interactions that define the native state, reaching the folded state

molten-globule states, with a small fraction of native states. Finally, when quenching for a few seconds, we only found natively folded proteins. This time evolution in the relative populations of the molten globule/native state suggests that the molten globule is an immature protein state that precedes the transition to the native state, in accord with the idea of a collapsed protein conformation, yet to establish the native contacts. Thus, we can model the folding pathway of protein L as a simple three-state kinetic model:

$$U \xrightarrow{r_{MG}} MG \xrightarrow{r_N} N$$

where r_{MG} is the kinetic rate to form the molten globule (MG), and r_N is the maturation rate from the molten globule state to the native one (N). By solving this kinetic model, we can work out the probability of reaching the molten globule state (P_{MG}) or the native state (P_N) as a function of the folding time Δt [7]:

$$P_U(t) = e^{-r_{MG}t}$$

$$P_{MG}(t) = \frac{r_{MG}}{r_{MG} - r_N}[e^{-r_N t} - e^{r_{MG}t}]$$

$$P_N(t) = \frac{r_{MG}}{r_{MG} - r_N}\left(1 - e^{-r_N t}\right) - \frac{r_{MG}r_N}{r_{MG} - r_N}te^{-r_{MG}t}$$

Fitting our experimental data to the kinetic model allowed us to extract the rates of formation of the molten globule ($r_{MG} = 10.97 \pm 1.42$ s^{-1}) and native state (r_N = 1.28 ± 0.24 s^{-1}). With this information, we can model the folding pathway of protein L as a two-step reaction, where the unfolded polypeptide first collapses to a molten-globule-like state over ~ 0.09 s, characterized by a low mechanical stability. From this immature folded state, and over a timescale of ~ 0.8 s, the protein reaches its native folded state (Fig. 12.3d).

12.3.2 Protein Folding Dynamics Under Complex Force Signals

Most force spectroscopy techniques rely on the application of simple force pertur-bations, typically force ramps, to test the mechanical stability of the studied protein or constant forces to monitor its equilibrium dynamics. However, mechanical cues inside the cell are unlikely to resemble such simple shapes; the cell environment is naturally noisy, and mechanical signals typically change quickly over time, so the ability to respond to such fast force fluctuations is a crucial capability of many biological systems [30–32]. The human auditory system is a formidable example in this context, capable of converting complex vibration patterns into electrophysiolog-ical signals with high sensitivity [32, 33]. Even more generally, cells are known to respond to force oscillations exerted by cyclic stretching of their substrate, and such complex stimuli can trigger mechanosignalling pathways to control cellular behavior [31]. Therefore, there exists a natural motivation to understand the response of indi-vidual proteins to complex mechanical signals, including mechanical noise and force oscillations.

The implementation of a magnetic tape head in our single-molecule magnetic tweezers unlocks the previous instrumental limitations, allowing us now to generate arbitrarily complex force signals that are directly applied to a tethered protein. To demonstrate our approach, we studied the dynamics of the talin R3 domain harboring the IVVI mutation, which increases its mechanical stability while maintaining its biological function [34, 35]. Talin is a key protein in focal adhesions, where it crosslinks active integrins with the actin filaments, regulating the engagement and maturation of the cell-substrate adhesion. In particular, the talin R3 domain is the weakest of the 13-rod domains and has been shown to regulate the cell's sensitivity to the substrate stiffness, likely by unfolding at low mechanical forces and recruiting vinculin to focal adhesions [35, 36].

Under constant forces between 8 and 10 pN and over ~ minute timescales, the talin R3^{IVVI} domain behaves like a classic two-state folder, transitioning stochastically between its folded and unfolded states with well-defined rates [11]. At 9 pN, it populates the folded and unfolded states with equal probability, giving rise to its characteristic "hopping" dynamics (Fig. 12.4a, left). The dwell times in the folded (or unfolded) state are distributed exponentially, as expected for a simple two-state folder that switches between its folded and unfolded states by overcoming a single free energy barrier (Fig. 12.4a, right).

However, when we subjected R3^{IVVI} to a force signal of the same average force (9 pN) but oscillating with a small amplitude (0.7 pN) at a frequency of 1 Hz, we observed a dramatic change in R3^{IVVI}'s dynamics. Under this small oscillation, the folding and unfolding transitions are mostly synchronous with the driving force signal, indicating that talin can detect this low-amplitude force oscillation (Fig. 12.4b,

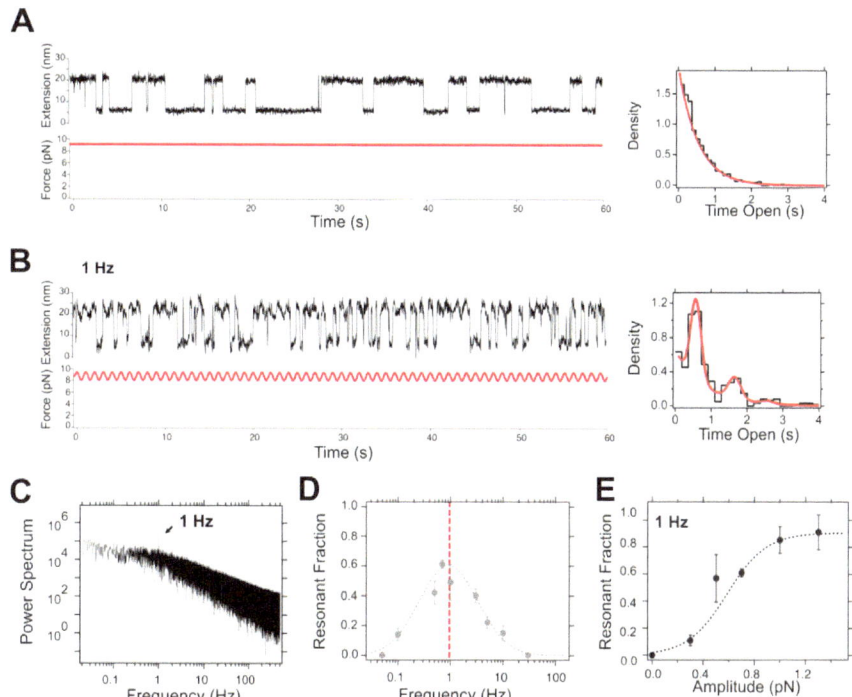

Fig. 12.4 Talin R3^{IVVI} responds to oscillatory force signals in a finely-tuned way. a (Left) Typical magnetic tweezers recording of R3^{IVVI} at 9 pN, where it populates the folded and unfolded states with equal probability. (Right) Distribution of dwell times in the open (unfolded) state. R3^{IVVI} transitions stochastically between the folded and unfolded state, giving rise to exponentially distributed dwell times. **b** (Left) Dynamics of R3^{IVVI} under an oscillatory force signal with a frequency of 1 Hz, an average value of 9 pN, and an amplitude of 0.7 pN. The folding and unfolding transitions get synchronized with the force oscillations. (Right) Distribution of dwell times in the unfolded state, which shows Gaussian peaks at 0.5, 1.5, and 2.5 s, arising from the synchronized transitions, and an underlying exponential contribution accounting for the remanent stochastic transitions. **c** Power spectrum of R3^{IVVI} folding dynamics under the 1 Hz force signal, indicating a clear resonant peak at the driving frequency. **d** Resonant fraction of R3^{IVVI} transitions as a function of the driving frequency for signals with an amplitude of 0.7 pN. Talin's response is strongly frequency-dependent, and it only detects signals oscillating at ~ 1 Hz, filtering out higher or lower frequencies. **e** Dependence of talin's response with the amplitude of the driving signal. Talin is highly sensitive, being able to detect signals as small as 0.3 pN

left). This change in behavior is reflected in the distribution of dwell times, now showing Gaussian-like peaks centered at odd multiples of half the driving frequency (0.5, 1.5, 2.5 s), with an underlying exponential contribution, reflecting a remanent stochastic behavior. By fitting the distribution of dwell times to a combination of Gaussian peaks and an exponential function [37], we can characterize the strength of talin's response to the force oscillations as the relative weight of the Gaussian peaks (entrained transitions) and exponential (stochastic transitions). Furthermore, calculating the power spectrum of talin's dynamics (removing the elastic contribution of the polypeptide chain, which follows the force oscillations), we observe a clear peak at the driving frequency of 1 Hz, indicative of resonant dynamics (Fig. 12.4c).

Similarly, we explored the dynamics of $R3^{IVVI}$ over a broad range of frequencies (spanning between 0.1 and 100 Hz) and characterized its response by the fraction of entrained transitions (resonant transitions). Figure 12.4d shows the resonant fraction as a function of the driving frequency for 0.7 pN amplitude signals. The peaked dependence, with an optimal response in the ~ 1 Hz range, reveals that $R3^{IVVI}$ is exquisitely sensitive to the oscillation frequency of the applied force. Very low frequencies are not detected as the force changes too slowly, and most transitions are stochastic; similarly, very high frequencies are rejected as talin folding/unfolding dynamics cannot follow such rapid force changes. Importantly, the resonant frequency is related to the natural folding/unfolding rates of talin at the coexistence force. As we demonstrated, proteins with lower mechanical stability and faster folding/unfolding rates synchronize their dynamics at higher frequencies of the timescale of the natural transition rates [38]. This behavior suggests that talin operates as a "mechanical bandpass filter," responding only to a narrow range of frequencies. When increasing the amplitude of the driving signal, we naturally increase talin's response (Fig. 12.4e); yet, talin is capable of detecting very weak signals of just 0.3 pN of amplitude.

This behavior is reminiscent of the physical phenomenon of stochastic resonance, by which nonlinear systems exhibit an amplified response to a weak input signal thanks to the presence of noise [37]. Stochastic resonance requires three basic ingredients: (1) a bistable system; (2) a weak periodic input; (3) an intrinsic source of noise. All these conditions are fulfilled in our experiment since talin folding dynamics can be well-described as a simple two-state system, we are applying a weak periodic force perturbation, and the thermal bath provides the intrinsic noise source.

Stochastic resonance has been described in a broad range of physical problems, from climate dynamics to quantum systems [37]. Similarly, many biological systems, mainly in the context of mechanoreceptors, have been shown to take advantage of noise as a method of signal detection [39, 40]. Our experiments clearly show that the talin R3 domain, a key protein mechanosensor, detects not only the magnitude of the force perturbation but also its frequency of oscillation, adding a new layer to its previously known force sensing capabilities. While it remains unclear if stochastic resonance is a general phenomenon among force-sensing proteins, our experiments are strongly suggestive that proteins with similar folding properties will be able to respond to force oscillations similarly, being the responsive frequency range selected by their mechanical stability. Therefore, it is enticing to speculate that mechanosensing proteins composed of tandem repeats of different domains with different mechanical stabilities will be responsive to a broad range of mechanical signals, where each domain will detect a specific frequency range, providing a nanomechanical mechanism to Fourier-decompose complex mechanical signals.

12.4   Conclusion and Outlook

While traditionally devoted to the study of nucleic acids, magnetic tweezers have become, over the past few years, an ideal technique for studying protein nanomechanics. This owes mainly to the development of new anchoring strategies—such as the HaloTag chemistry—and instrumental developments that increased the stability and resolution of initial designs. In this chapter, we have focused on the improvement of the force-application strategy by implementing a magnetic tape head that substitutes the more usual permanent magnet approach. The ability to manipulate force through electric current provides a faster and more stable way of subjecting proteins to force in a highly controlled manner. This is particularly critical for the study of proteins with very low mechanical stability, such as those relevant in mechanotransduction, which are exposed to forces that rarely exceed 10 pN, being thus exquisitely sensitive to meager force changes [41]. Furthermore, the possibility of changing the force very quickly opens the gates to interrogating protein dynamics under novel force protocols, which could potentially undercover new biophysical phenomena, as demonstrated in the case of talin dynamics under force oscillations. An additional advantage of magnetic tweezers not discussed here is its great stability, which permits measuring protein dynamics over very long timescales, up to several hours or even days [10, 11]. Similarly, this opens up the possibility of exploring new questions in protein folding, such as molecular heterogeneity or the effect of damaging posttranslational modifications triggered by protein aging.

Still, there are several directions to further develop magnetic tweezers instrumentation and overcome some of its limitations. A disadvantage of magnetic tweezers compared to optical tweezers or the AFM is its lower temporal resolution, which owes to the slow image acquisition and analysis methods compared to laser-based detection. While current designs reach sampling rates up to 1.5 kHz, the implementation of faster cameras and new computing methods, such as FPGAs, could potentially increase the time resolution to capture μs-timescales. Likewise, the bespoke design of most magnetic tweezers models enables the easy implementation of novel instrumental capabilities, among which the single-molecule fluorescence is likely the most appealing one. Combined fluoresce-magnetic tweezers instruments have been previously demonstrated for the study of protein-DNA interactions [42], where a double-stranded DNA molecule is tethered to the magnetic bead and mechanically stretched while binding fluorescently-labeled proteins are detectable with, for example, TIRF microscopy, allowing to correlate binding reactions with molecular nanomechanics. In this sense, a natural evolution of the current magnetic tape-head implementation is to include fluorescence capabilities for the study of protein–protein interactions under force.

References

1. P. Roca-Cusachs, V. Conte, X. Trepat, Nat. Cell Biol. **19**, 742–751 (2017)
2. E.C. Yusko, C.L. Asbury, Mol. Biol. Cel **25**, 3717–3725 (2014)
3. K.C. Neuman, A. Nagy, Nat. Meth. **5**, 491–505 (2008)
4. M. Mora, A. Stannard, S. Garcia-Manyes, Chem. Soc. Rev. **49**, 6816 (2020)
5. C. Bustamante et al., Annu. Rev. Biochem. **89**, 443–470 (2020)
6. I.D. Vlaminck, C. Dekker, Annu. Rev. Biophys. **41**, 453–472 (2012)
7. R. Tapia-Rojo, E.C. Eckels, J.M. Fernandez, Proc. Natl. Acad. Sci. U.S.A. **116**, 7873–7878 (2019)
8. I. Popa et al., J. Am. Chem. Soc. **138**, 10546–10553 (2016)
9. A. Lof et al., Proc. Natl. Acad. Sci. U.S.A. **116**, 18798–18807 (2019)
10. R. Tapia-Rojo, et al., Biophys. J. **123**, 814–823 (2024)
11. R. Tapia-Rojo et al., Nat. Phys. **19**, 52–60 (2023)
12. A. Alonso-Caballero, et al., bioRxiv 2021.01.04.425265 (2021)
13. K.H.L. Buschow, G.J. Long, F. Grandjen, *High Density Digital Recording* (Kluwer Academic Publishers, Norwell, MA, 1993)
14. I. Popa et al., J. Am. Chem. Soc. **135**, 12762–12771 (2013)
15. A. Alonso-Caballero et al., Nat. Chem. **13**, 172–181 (2021)
16. Z. Yu et al., Rev Sci. Instr. **85**, 123114 (2014)
17. J. Lipfert, X. Hao, N.H. Dekker, Biophys. J. **96**, 10546–11553 (2009)
18. O. Karlqvist, *Transactions Royal Institute Technology* (Henrik Lindstahls Bokhandel, Stockholm, 1954)
19. R. Lawrence Comstock, *Introduction to Magnetism and Magnetic Recording* (Wiley-Interscience, 1999)
20. G. Fonnum, et al., J. Mag. Mag. Mater. **293**, 41–7 (2005)
21. M. Doi, S. Edwards, *The Theory of Polymer Dynamics (*Oxford Science Publications, Oxford, 1986)
22. J. Valle-Orero et al., J. Phys. Chem. Lett. **8**, 3642–3647 (2017)
23. A. Sali et al., Nature **369**, 248–251 (1994)
24. J. Jumper et al., Nature **596**, 583–589 (2021)
25. K.A. Dill, H.S. Chan, Nat. Struct. Biol. **4**, 10–19 (1997)
26. V.S. Pande, D.S. Rokshar, Proc. Natl. Acad. Sci. U.S.A. **95**, 1490–1494 (1998)
27. C. Cecconi et al., Science **309**, 2057–2060 (2005)
28. S. Garcia-Manyes, et al., Proc. Natl. Acad. Sci. U.S.A. **106**, 10534–10539 (2009)
29. D.E. Kim, C. Fisher, D. Baker, J. Mol. Biol. **298**, 971–984 (2000)
30. H. Johnson, Q. Rev, Biol. **62**, 141–152 (1987)
31. S.V. Plotnikov et al., Cell **151**, 1513–1527 (2012)
32. A. Hudspeth et al., Neuron **59**, 530–545 (2008)
33. M.A. Vollrath et al., Annu. Rev. Neurosci. **30**, 33–365 (2007)
34. B.T. Goult et al., J. Biol. Chem. **288**, 8238–8249 (2013)
35. R Tapia-Rojo, A. Alonso-Caballero, J.M. Fernandez, Sci. Adv. **6**, eaaz4707 (2020)
36. A. Elosegui-Artola et al., Nat. Cell Bio. **18**, 540–548 (2016)
37. L. Gammaitoni et al., Rev. Mod. Phys. **70**, 223–287 (1998)
38. R. Tapia-Rojo, A. Alonso-Caballero, J.M. Fernandez, Proc. Natl. Acad. Sci. U.S.A. **117**, 21346–21353 (2020)
39. J.K. Douglass et al., Nature **365**, 337–340 (1993)
40. J.E. Levin et al., Nature **380**, 165–168 (1996)
41. P. Ringer et al., Nat. Meth. **14**, 1090–1096 (2017)
42. F.E. Kemmerich et al., Nano Lett. **16**, 381–386 (2016)

Chapter 13
Design of Iron Oxide Nanoparticles as Theranostic Nanoplatforms for Cancer Treatment

Thomas Gevart, Barbara Freis, Thomas Vangijzegem, Maria Los Angeles Ramirez, Dimitri Stanicki, Sylvie Begin, and Sophie Laurent

13.1 Iron Oxide Nanoparticles for MRI

13.1.1 Synthesis of IONPs

Current challenges in the synthesis of IONPs are to obtain controlled nanoparticles in terms of composition, size, shape, and crystallinity by avoiding undesired reactions and the use of too many reactants [1]. The specifications on the design of IONPs for combining dual treatment via magnetic hyperthermia lead to investigate the effect of

T. Gevart · B. Freis · T. Vangijzegem · D. Stanicki · S. Laurent (✉)
NMR and Molecular Imaging Laboratory, Department of General, Organic and Biomedical
Chemistry, UMONS, 19 Avenue Maistriau, 7000 Mons, Belgium
e-mail: sophie.laurent@umons.ac.be

T. Gevart
e-mail: thomas.gevart@umons.ac.be

B. Freis
e-mail: barbara.freis@ipcms.unistra.fr

T. Vangijzegem
e-mail: thomas.vangijzegem@umons.ac.be

D. Stanicki
e-mail: dimitri.stanicki@umons.ac.be

B. Freis · M. L. A. Ramirez · S. Begin
Institut de Physique et Chimie des Matériaux, UMR CNRS-UdS 7504, Université de Strasbourg,
CNRS, 23 Rue du Loess, BP 43, 67034 Strasbourg, France
e-mail: mdlaramirez@unistra.fr

S. Begin
e-mail: sylvie.begin@ipcms.unistra.fr

© The Author(s) 2024   175
B. Doudin et al. (eds.), *Magnetic Microhydrodynamics*, Topics in Applied Physics 120,
https://doi.org/10.1007/978-3-031-58376-6_13

the size and shape of IONPs. A synthesis method allowing easy modulation of the size and shape of IONPs and leading to a narrow size distribution and high colloidal stability is needed.

90% of superparamagnetic IONPs are synthesized by chemical methods according to Ali et al. [2], even if physical and biological methods keep showing great interest [2, 3]. Among the chemical methods, coprecipitation, a method developed by Massart et al. is the most used one [4]. Iron (II) and iron (III) salts are dissolved in an aqueous solution and are precipitated after the addition of a base through this global reaction:

$$Fe^{2+} + 2\,Fe^{3+} + 8\,OH^- \rightarrow Fe_3O_4 + 4\,H_2O \tag{13.1}$$

The main advantage of this method is the production of IONPs with a high yield directly in water with an easy-to-process and cheap method. However, if no ligand is added during the synthesis, NPs tend to form aggregates, as they are 'naked'. Nonetheless, the addition of ligand during the process can disrupt the formation of NPs and lead to NPs with no good control of size or shape [5].

Thus, through the years, other methods have been developed to improve the control of size, shape, and colloidal stability of the synthesized nanoparticles. Among them, microemulsion, polyol synthesis, hydrothermal, or microwave-assisted synthesis can be cited [1]. Another growing method since the early 2000s is thermal decomposition (TD). Most reported methods are summarized Table 13.1 with their ability to control the size and shape of IONPs [1, 6].

The thermal decomposition synthesis method is generally designated as the most suitable one to obtain particles with controlled size and shape as well as colloidal stability. First introduced by Hyeon's and Sun's research groups [7, 8], this method rapidly gathered interest among researchers because of its various advantages. Indeed, this method allows synthesizing IONPs with a high yield. The NPs are also coated in situ with a surfactant, thus do not agglomerate, and present high monodispersity in size. Moreover, the synthesis allows precise control of both particle size

Table 13.1 Size and shape control of the most reported synthesis methods of iron oxide NPs

Method	Process conditions (temperature, process time, handling)	Dispersity	Shape control	Yield
Co-precipitation	Low T°, minutes, complicated ++	Average	Bad (irregular sphere)	High
Microemulsion	Low T°, hours, complicated	± narrow	Very good (cube-sphere)	Poor
Sol–gel	Low T°, hours, simple	± narrow	Good (sphere)	Poor
Hydrothermal	High T°, hours, simple	± narrow	Bad (irregular sphere)	High
Thermal decomposition	High T°, hours-days, complicated ++	Very narrow	Very good (cube-sphere)	High

Adapted from [1]

and shape, which can be tuned by changing synthesis parameters such as the nature of reactants, the nature of solvents, the heating rate or the reaction duration. However, the synthesis requires quite harsh conditions, as the mixture needs to be heated at a high temperature for several hours in a high boiling point organic solvent. The surfactants are generally hydrophobic and ligand exchange is necessary to make the NPs stable in water and physiological media for further biological applications.

13.1.1.1   Principles of the TD Method

The TD method was at first used for the synthesis of quantum dots and semi-conductor nanocrystals in the late 1990s [9, 10]. An organometallic complex is brought to high temperature in a high boiling point organic solvent and in the presence of a surfactant to stabilize the formed NPs (Fig. 13.1a). The TD process leads to the decomposition of the precursor and the generation of monomers, which then trigger the formation of small nuclei. Then, depending on the reaction temperature and time, these nuclei will next grow into well-crystallized NPs stabilized by the surfactant coating (Fig. 13.1b). This method allows a good separation between the nucleation and growth steps, which is a prerequisite to control the size distribution. Indeed, it follows the LaMer and Dinegar theory of nucleation and growth of particles, introduced in 1950 to explain the formation of monodisperse hydrosols [11]. This theory is based on three different steps depending on the concentration of monomers in solution and it is generally applied to the mechanism of NPs formation via TD [12].

Three major stages are thus proposed for the IONPs synthesis (Fig. 13.1):

(i)   iron-based monomer generation (monomers are reported to result from the precursor decomposition upon increasing temperature) [13].

(ii)   nucleation after which a critical nucleation concentration (C_{min}^{nu}) in monomer is reached.

(iii) growth of nuclei after which the monomer concentration falls below C_{min}^{nu} but stays above the saturation C_s. Therefore, nuclei are generated during a nucleation step that is followed by a homogeneous growth step without the creation of new nuclei [13–15].

LaMer theory can thus explain NPs synthesis with a narrow size distribution and the possibility to obtain different shapes using shape-driving ligands. Nevertheless, a continuous growth process from monomer has also been proposed [16, 17] and a recent study reported that instead of a homogeneous nucleation within the solvent, the nucleation occurs within vesicle-like "nanoreactors" which confine the reactants [18]. The nature of ligand and solvent used also have an impact on the stability of the precursor and its decomposition process [14]. The TD synthesis with its variety of experimental parameters such as temperature, reaction time, concentration and nature of precursors, surfactants and solvent offers great freedom in the design of NPs (e.g. to tune the size, morphology and composition).

IONPs synthesis by the TD method has been developed since the early 2000s [7, 8, 19]. The most common iron precursors are $Fe(acac)_n$ (acac = acetylacetonate)

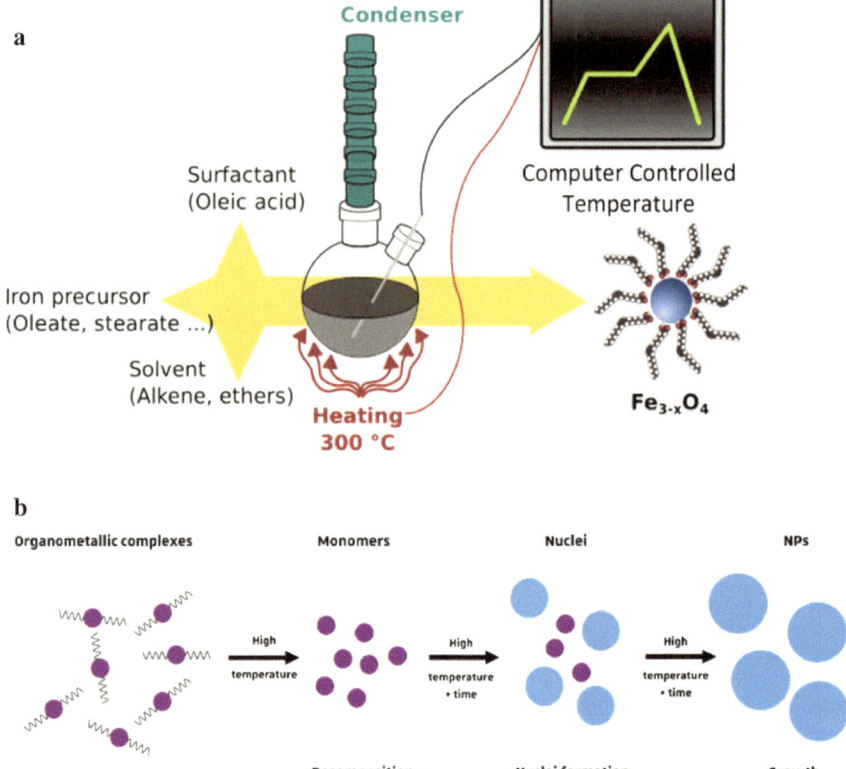

Fig. 13.1 **a** Schematic representation of the thermal decomposition method, and **b** The three major steps of the TD synthesis

[8, 20, 21] and iron oleate [19, 22–27] (unsaturated C18) but other precursors are also used such as iron stearate [14, 16, 28–31] (saturated C18) or carbonyls $Fe(CO)_x$ [32, 33]. The boiling point of organic solvents must be quite high to decompose the precursor and is generally in the range 270–350 °C. For this reason, mainly alkenes and ethers are used [14, 22, 34]. Finally, commonly used surfactants are fatty acids like oleic acid (OA) and sometimes mixtures of fatty acids [35]. Their role is crucial as they stabilize the formed NPs and prevent them from aggregation by Van der Walls and dipolar interactions.

Two ways of performing the thermal decomposition of iron precursors are possible. The first one is the 'hot injection' method where the iron precursor is directly injected into the solvent and surfactant mixture, already heated at high temperature [36]. The iron precursor will directly decompose. However, this quite abrupt method presents some issues for shape control and it is not suited for scale-up [37]. The second one is the 'heating-up' method, where the precursor is solubilized with the surfactant in the organic solvent at ambient temperature and the mixture is heated up to the boiling temperature of the solvent. This method allows more control of the decomposition [38].

By tuning various parameters of the TD method such as the nature of solvent, of precursors and of the surfactants, the amount and the number of surfactants, the concentration of reactants but also the reaction duration, IONPs with different sizes and shapes may be synthesized [39]. The strength of this method comes from the observed separation between the nucleation stage and the growth stage. However, as the parameters influencing NPs size, shape and composition are completely linked and entangled together, it is still difficult to fully understand the process and to predict exactly how a parameter influences the final synthesized NPs.

To obtain nanoparticles with a diameter higher than 15 nm, the synthesis needs to ensure a longer growth step or to favor a low yield in nuclei during the nucleation step preserving monomers for the growth step. The main parameters reported to tune the size of IONPs are the nature of the solvent, the reaction temperature and the surfactant to iron precursor ratio [14, 22, 40, 41]. However, an impact of the solvent, precursor and reaction time have also been evidenced [14].

- **Reaction temperature tuned by the nature of solvents**

The effect of the solvent is mostly related to its boiling point; the higher it is, the larger the NPs diameter should be. Indeed, a higher boiling point and a longer growth step should yield more monomers (as some iron precursors decompose on large temperature range). However, the nature of the solvent (e.g. its polarity) may also affect the NPs size by affecting the precursor's thermal stability and thus its decomposition [14]. Highly polar organic solvents would favor the decomposition of a large amount of precursor, inducing germination of small nuclei with less monomer available for the grain growth step: small-sized NPs are thus obtained.

Non-polar solvents such as alkenes do not interact with iron precursors and so the diameter increases in a quite linear way with the solvent boiling point. This confirms that the growth rate mainly depends on the reaction temperature. In that context, docosene is suitable to synthesize IONPs with a mean size around 20 nm. However, docosene is solid at ambient temperature (its melting temperature is at 62 °C) which makes the washing and purification steps difficult. Some groups prefer to work with an alkane solvent squalane, which is not solid at ambient temperature and presents a high boiling point of 470 °C [39, 42, 43]. However, as its boiling point is so high, we cannot use it at the boiling point to avoid degradation of the alkyl chains of the reactants. Thus, the temperature has to be kept below the boiling point where it is less controlled. It was observed that depending on the effective reaction temperature, some variations in the composition from a spinel phase to core–shell NPs with a wüstite core occur [34]. A low heating rate during the growth step was shown to be promising to ensure a spinel composition [34]. The addition of a small amount of dibenzyl ether (DBE) with octadecene or squalane solvent would provide a more oxidative environment and allow IONPs with mean size higher than 15 nm and a spinel composition to be obtained [43–47].

- **Influence of the molar ligand/precursor ratio**

Numerous studies have been conducted on this parameter; the influence of the amount of ligand on the NPs size is quite complex [14, 40, 41, 48], either an increase [49, 50] or the opposite trend [51] was observed. Bronstein et al. [40] observed an evolution with iron oleate and oleic acid in eicosane without proposing an explanation, while Salas et al. [41] reported an influence of oleic acid on nucleation and growth rates. Indeed, two competitive mechanisms may occur. First a higher stabilization of the iron complex, with the increased amount of oleic acid which thus decomposes at higher temperature. But also, a stronger stabilization of nuclei affecting the grain growth. Thus, this parameter appears more complex to tune to control the growth of IONPs.

- **Reaction time**

Another obvious parameter to increase the size of the NPs is the reaction time. Baaziz et al. made experiments on the established 10 nm protocol and concluded that an increase in reaction time leads to an increase of size from 10 nm up to 14/15 nm after 6 hours [14]. Adapting the reaction time with solvents with higher boiling point could be a way to reach sizes around 20 nm.

- **Effect of the nature of precursors**

Hufschmid et al. [52] also demonstrated that some precursors are more suited for IONPs synthesis of specific size ranges. Bronstein et al. evidenced an effect of washing and aging conditions of iron oleate [48] and alkyl chain length [40] on the IONPs size and shape. Recent research work has been conducted to improve the quality and stability of the iron oleate precursor (reported highly sensitive to minor variations in its synthesis due to its propensity to retain water, oleic acid, and other reaction by-products) by performing different washing or drying treatments. These treatments were found efficient to better stabilize the iron oleate precursor and obtain IONPs with larger sizes, up to 40 nm [45, 53, 54].

The investigation of the thermal decomposition of both iron stearate $FeSt_2$ and $FeSt_3$ precursors in standard synthesis conditions of 10 nm spherical NPs (dioctylether as solvent and oleic acid as surfactant) led to spherical NPs with a monocrystalline structure and a homogeneous $Fe_{3-x}O_4$ composition; only the size was slightly affected by the nature of iron stearate (about 9 nm with $FeSt_3$ and 10 nm with $FeSt_2$) [34]. This was attributed to the fact that the decomposition kinetics of $FeSt_2$ at temperature below 298 °C was higher than that of $FeSt_3$, which decomposes in a larger temperature range up to 350 °C. When the TD experiments occur in solvent or mixture of solvents with higher boiling point to obtain IONPs with higher size, an impact of iron stearates was observed. Only with $FeSt_3$, IONPs with sizes higher than 15 nm were obtained. When $FeSt_2$ is used as a precursor, an increase of the size is observed, but it stays lower than 15 nm. Indeed, $FeSt_2$ decomposes mainly below 300 °C and only few precursors are available for the growth step by contrast with $FeSt_3$, which decomposes up to 350 °C and provides monomers during the growth step [34, 55].

13.1.1.2   IONPs Composition

At the nanoscale, oxidation phenomenon becomes much more important than at the bulk scale. Indeed, at the nanoscale, the iron (II) cations in magnetite $(Fe^{3+})_A[Fe^{2+}Fe^{3+}]_BO_4{}^2$ (where A represents the cations in the tetrahedral sites and B those in the octahedral sites) become highly sensitive to oxidation especially those located at the surface of the NP [14, 56–59].This Fe^{2+} oxidation is very sensitive to the IONPs size and thus, a IONPs size dependent composition was observed in nanoparticles synthesized by coprecipitation [57, 60] and by thermal decomposition [14]. In the case of the thermal decomposition method, Baaziz et al. [14] have shown that when NPs have a diameter smaller than 8 nm, the oxidation phenomenon of Fe(II) cation into Fe(III) cation is quite total. So, the NPs have a composition very close to that of the maghemite phase. For IONPs with a diameter higher than 12 nm, the oxidation of Fe(II) cations takes place mainly at their surface. Thus, the IONPs are composed of a magnetite core and an oxidized layer at the surface ($Fe_{3-X}O_4$) and the higher is the NP size, the higher is the size of the magnetite core. When the diameter is intermediate (this means above 8 nm but below 12 nm), the magnetite phase is partially oxidized without the appearance of a core–shell structure. Thus, the overall composition is described as being $Fe_{3-X}O_4$. Further Mössbauer spectrometry characterizations have shown the presence of oxidation defects, surface and volume spin canting as a function of NPs diameter [57].

Small NPs presented mainly a surface spin canting. NPs with larger sizes display different oxidized shell thickness, defects and surface spin canting. NPs with intermediate sizes display a surface and in particular a volume spin canting due to a disordered structure induced by a perturbed oxidation state in these NPs [61].

During the synthesis of IONPs larger than 15 nm [23, 34, 39] or with different shapes such as the cubic form [25, 26, 62], the formation of core–shell structures with a wüstite core and a spinel shell ($Fe_{1-X}O@Fe_{3-X}O_4$) has often been reported. One main reason is that the nuclei formed during the TD process have a wüstite composition. To increase the size of IONPs, the synthesis is performed at higher temperature to favor the growth step and organic solvents with higher boiling point are used. These solvents are often non-polar solvents such as squalane or octadecene providing a reducing environment which is not favorable for oxygen diffusion.

For the standard synthesis of 10 nm IONPs, the final composition of IONPs corresponds to oxidized magnetite $Fe_{3-X}O_4$. This suggests that the oxidation of nuclei has occurred simultaneously with their growth and this is in agreement with the fact that the nucleation occurs at about 280 °C, a temperature which is very close to the boiling point of octylether (290 °C) [13, 18, 31, 34, 48]. Octylether would not provide a reducing environment by contrast with alkene solvents and the heating time of 2 h at this boiling point should favor the oxidation of the wüstite nuclei. In fact, problems arise when trying to synthesize IONPs with sizes higher than 15 nm [14, 34]. Indeed, if the oxidation kinetics is too slow compared to the growth kinetics, which is an issue often appearing, core–shell structures are obtained (Fig. 13.2). If an iron oxide spinel structure is identified, the IONPs often contain defects [14, 23, 34, 63]. It has often been reported that the oxidation of wüstite induces the presence of defects such as

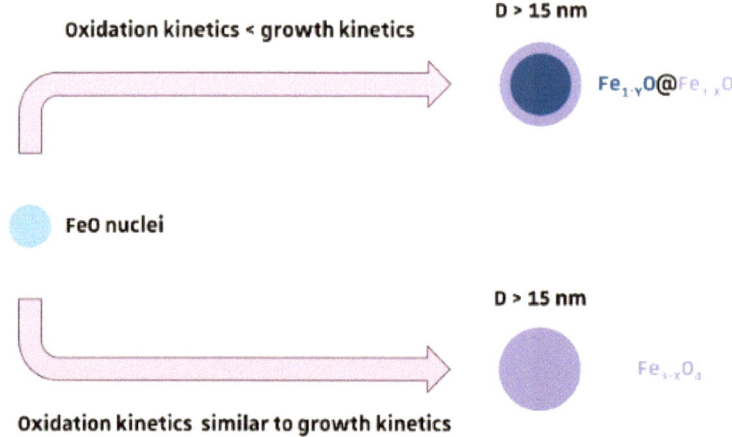

Fig. 13.2 Two possible compositions of big-sized IONPs (diameter > 15 nm) depending on the oxidation rate

dislocations or antiphase boundaries observed as function of the IONPs size [23, 25, 28, 34, 39, 44–46, 63]. Such defects have a high impact on the magnetic properties of IONPs and thus on their magnetic hyperthermia performances [64, 65].

The solvent nature, a too dense surfactant layer at the surface of nuclei, a high heating rate are among reported parameters hampering a good oxidation kinetics. More and more studies demonstrated that the addition of DBE is a good way to avoid the formation of core–shell IONPs [43–47]. Indeed, the solvent's redox activity would be very important to control the valence state of iron. Thermolysis of aromatic ethers produces oxidizing species that stabilize the inverse spinel phase, while alkene hydrocarbons have reducing effects which can favor the formation of wüstite. Controlling this non-aqueous redox environment enables reproducible and scalable synthesis of nearly defect-free IONPs in the 10–30 nm range without the need for post-synthesis modification [45–47, 66, 67]. This "non-aqueous redox phase-tuning" method is a very suitable method to avoid the formation of the wüstite phase during the nanoparticle growth process. Indeed, redox active species, coming from alkene solvents such as 1-octadecene and from ether solvents such as DBE, are generated during the high temperature synthesis stage. Specifically, DBE decomposition generates benzaldehyde, which possesses oxidative character [47]. On the contrary, the tendency of the 1-octadecene's vinyl group to oxidation produces a reductive effect.

The competition between growth and oxidation depends greatly on the synthesis methods and the reactants present during synthesis. The oxidation of the wüstite core during synthesis can also lead to various amount of structural defect in a NP. Moreover, this core–shell structure and the presence of defects can greatly alter the magnetic properties.

Table 13.2 Magnetic behavior of the wüstite, the maghemite and the magnetite phases at the bulk scale

Types of magnetism	Susceptibility χ	Magnetic moment μ and atomic behavior	
Ferri magnetism Fe$_3$O$_4$, γ-Fe$_2$O$_3$ Phases	Large and positive, function of the applied field H	Anti-parallel aligned magnetic moments without compensation	
Antiferromagnetism FeO phase	Small and positive	Anti-parallel aligned magnetic moments with compensation	

13.1.2   Magnetic Properties of IONPs

13.1.2.1   Magnetic Properties of Bulk Magnetite and Maghemite

Magnetic Properties of Bulk Magnetite, Maghemite and Wüstite

Magnetic materials are described thanks to three main parameters: their magnetic moment μ corresponding to the tendency of their dipoles to align under an external magnetic field H, their magnetization M corresponding to the magnetic moment per volume, and their magnetic susceptibility χ corresponding to the propensity of the material to align its magnetization M to the external magnetic field H. The relation between magnetization and the external magnetic field is given by:

$$M = \chi H \tag{13.2}$$

From the magnetization can be extracted the maximum of magnetization called saturation magnetization which is an important criterion to distinguish magnetic materials.

The wüstite, magnetite, and maghemite phases present different types of magnetism. The wüstite phase presents antiferromagnetic properties [68], which means that it has no magnetism without any field applied, and low magnetic properties when one is applied (Table 13.2). Maghemite and magnetite are both ferrimagnetic below their Curie temperature (858 K for magnetite [69] and 890 K for maghemite [70]) and thus present a spontaneous magnetic spin organization with an anti-parallel alignment and without compensation of the moment (Table 13.2).

The magnetic ordering disappears above the Curie temperature for ferrimagnetic compounds. The thermal agitation is sufficient to suppress spontaneous magnetization, and the compound becomes paramagnetic. This means that there is no magnetic order anymore without any field applied.

Ferrimagnetism of the magnetite phase is due to the presence of both iron (II) and iron (III) in octahedral and tetrahedral sites as shown in Table 13.3. The magnetic

Table 13.3 Magnetic structure, magnetic moment, and saturation magnetization of magnetite and maghemite

Iron oxides	Cations	Octahedral sites	Tetrahedral sites	Magnetic moment	Magnetization saturation M_s (Am²/kg)
Magnetite	Fe^{3+}	↑↑↑↑↑ Electronic transfer $5\,\mu_B$	↓↓↓↓↓ $-5\,\mu_B$	Cancellation	92
	Fe^{2+}	↑↓↑↑↑↑ $4\,\mu_B$		$4\,\mu_B$	
Maghemite	Fe^{3+}	↑↑↑↑↑ $\frac{5}{3}*5\,\mu_B$	↓↓↓↓↓ $-5\,\mu_B$	$3.33\,\mu_B$	74

moment of iron (III) in octahedral and tetrahedral sites can compensate. So, the magnetic moment of magnetite is determined by the moment of iron (II) in the octahedral site and is thus equal to 4 Bohr magneton (μB) which is a constant corresponding to the moment of an electron. The saturation magnetization of bulk magnetite is estimated to be 92 Am²/kg. For the maghemite phase, there are only iron (III) cations that have a moment of 5 μ_B in octahedral sites and tetrahedral sites. However, as vacancies also occupy octahedral sites, there are only 5/3 of iron (III) in octahedral sites which gives an overall moment equal to $\left(\frac{5}{3} - 1\right) \times 5\mu_B$ leading to a saturation magnetization of 74 Am²/kg.

At the bulk scale, ferrimagnetic materials present a magnetic structure made of domains called Weiss domains and separated by Bloch walls. This structure aims to diminish the internal energy of the material [68]. The magnetization is uniform within each domain but varies from one domain to another so that in the absence of an external magnetic field, there is no global magnetization.

When exposed to a magnetic field, the walls are moving so that domains with the same orientation as the applied field extend (Fig. 13.3). Structural defects can slow down this effect, which implies a delay in the magnetic response of the material. After this first exposure to a magnetic field, the material is always magnetized and keeps a permanent magnetization.

This phenomenon is at the origin of the hysteresis loop observed on the magnetization curve with a remanent magnetization M_R, a saturation magnetization M_S and a coercive field H_C characteristic of the studied material.

Anisotropy Energy

The magnetization vector in a ferrimagnetic crystal is not isotropic, the direction of the spins with respect to the crystal lattice depends on the magnetocrystalline

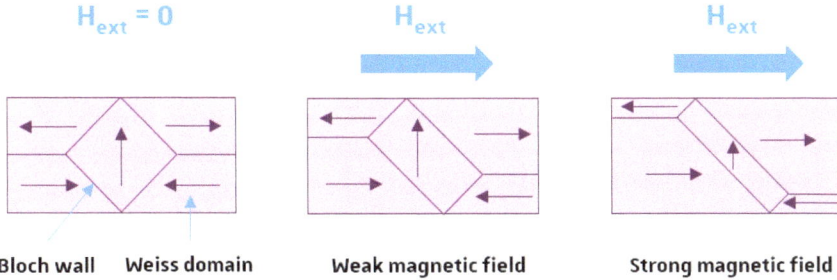

$H_{ext} = 0$ H_{ext} H_{ext}

Bloch wall   Weiss domain   Weak magnetic field   Strong magnetic field

Fig. 13.3 Creation of Weiss domain to diminish internal energy and behavior under a magnetic field

anisotropy. This anisotropy comes from the existence of what is called an "easy magnetization" axis within a crystal. This direction is induced by the crystallographic structure of the material studied. In the case of magnetite this easy magnetization axis is along the $< 111 >$ direction [68]. The magnetocrystalline anisotropy energy E_{MC} corresponds to the energy needed to reverse the magnetization from this easy axis.

Another type of anisotropy that can appear in ferrimagnetic materials is the spontaneous orientation with respect to the shape of the material, it is the shape anisotropy. In any non-spherical system, the magnetization tends to align itself along the largest dimension. This energy is proportional to the square of the saturation magnetization M_S, it is often dominant compared to the other sources of anisotropy. It imposes the direction of the magnetization at equilibrium in the absence of an external field [68].

13.1.2.2 Effect of the Nanoscale on the Magnetic Properties of Iron Oxides

Superparamagnetic Behavior of Magnetite and Maghemite at the Nanoscale

If the size of the ferrimagnetic material decreases under a critical diameter D_C, creating Bloch walls requires too much energy compared to having a single domain called monodomain. The monodomain exhibits oriented magnetization along its easy magnetization axis at room temperature. We speak of "monodomain blocked IONP". The spins inside such a blocked monodomain approach a single macrospin. Monodomain blocked IONPs display a wider opening of the hysteresis loop compared to ones with Weiss domains; they also present a remanent magnetization. The critical diameter for NPs composed of maghemite or magnetite is between 100 and 200 nm depending on the synthesis conditions. Esterlich et al. [71] reported diameter of 128 nm for magnetite and of 166 nm for maghemite (Fig. 13.4).

The macrospin of NPs within monodomain has two stable positions along the easy magnetization axis (parallel and antiparallel) as depicted in Fig. 13.5. They can switch their magnetization from one direction to the other by a movement called Néel

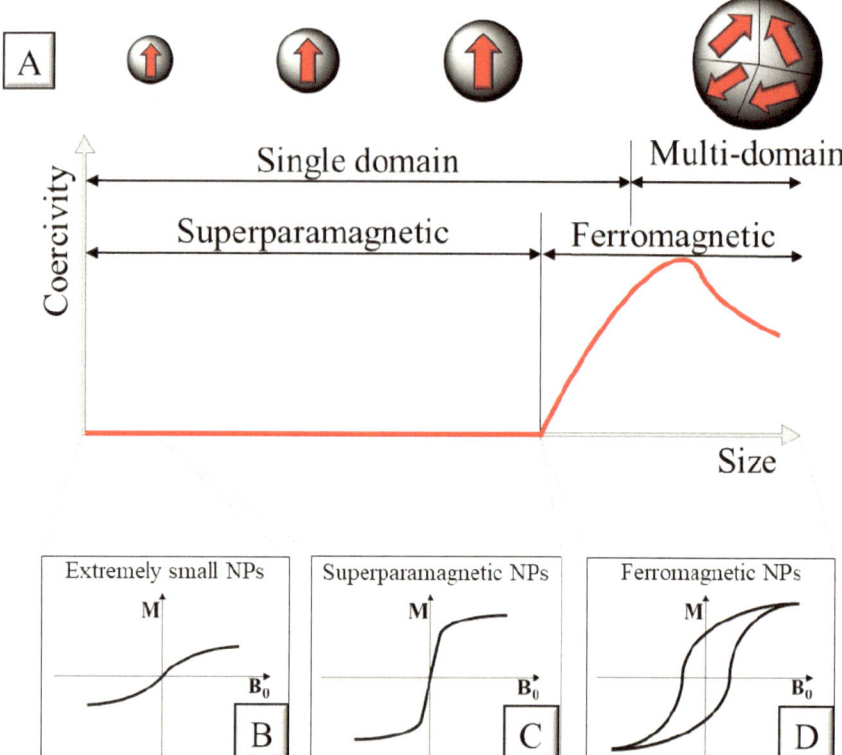

Fig. 13.4 (A) Evolution of coercivity as a function of the nanoparticle size. Coercivity decreases with the size of ferromagnetic materials until reaching the superparamagnetic state corresponding to single-domain nanoparticles with zero coercivity; (B–D) Magnetization curves of magnetic nanoparticles; (B) Extremely small NPs exhibit almost linear curve similar to paramagnetic substances due to very low magnetic properties; (C) Typical superparamagnetic curve of magnetic NPs with no remanence and coercivity; (D) Typical curve of ferromagnetic NPs exhibiting "magnetic memory". Adapted from [72]

relaxation that requires overcoming the anisotropy energy barrier KV where K is the anisotropy constant and V is the volume of the NP. For non-interacting magnetic NPs (no dipolar interactions), the probability that the magnetization spontaneously switches from one position to the other at a given temperature follows an Arrhenius law. Thus, the characteristic time for the Néel relaxation time τ_N is as follows:

$$\tau_N = \tau_0 e^{-\frac{KV}{k_B T}} \tag{13.3}$$

Here τ_0 is a length of time characteristic of the studied material, k_B is the Boltzmann constant, and T is the temperature.

When the size of the material is decreased again under a critical diameter D_{SPM}, the anisotropy energy KV becomes lower than the thermal energy $k_B T$ (Fig. 13.5a).

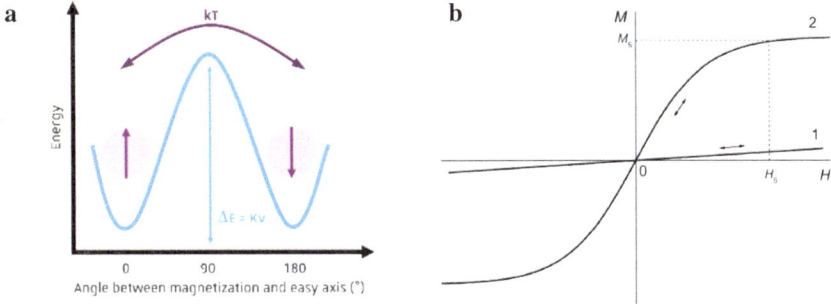

Fig. 13.5 **a** Magnetization flip between parallel and antiparallel orientations when magnetocrystalline anisotropy energy is comparable to thermal energy; **b** Schematic representation of the magnetization curve of a paramagnetic (1) material and a superparamagnetic (2) material

The critical diameter D_{SPM} value for magnetite and maghemite depends on their synthesis method but is reported to be around 20–25 nm for magnetite and 25 nm to 30 nm for maghemite [71].

As the anisotropy energy KV is lower than the thermal energy kT, magnetization can flip randomly from the two positions along the easy magnetization axis without any magnetic field applied. The material does not possess anymore a remanent magnetization at ambient temperature and the hysteresis loop is closed (Fig. 13.5b). The IONPs display thus a superparamagnetic behavior. The fact that superparamagnetic materials do not possess any magnetization when no magnetic field is applied is a key point for their use in biological applications. Indeed, it means that a suspension of superparamagnetic NPs won't aggregate due to magnetic interactions and can have good colloidal stability when injected in vivo.

When an external field is applied, the superparamagnetic material will align its magnetic moment with the field axis faster than a ferrimagnetic material as no Bloch walls are present. As soon as the external field is removed, the moments go back to a random position, giving no rise to a remanent magnetization or coercive field. Thus, when no magnetic field is applied, the material behaves like a paramagnetic material but has a high susceptibility.

Inversely, for superparamagnetic material, when the temperature is decreased below room temperature, there is a temperature from which the thermal energy becomes once again, lower than the anisotropy energy. This is the blocking temperature T_B, which is defined as follows (with τ_M the measurement time).

$$T_B = \frac{KV}{k_B \ln\left(\frac{\tau_M}{\tau_0}\right)} \tag{13.4}$$

So, under this blocking temperature T_B, the relaxation time will be higher than the measurement time thus, the magnetization will be blocked, with a magnetic moment rigidly oriented along the axis of easy magnetization. It must be noted that the blocking temperature depends on the size, the strength of the dipolar interactions,

Fig. 13.6 Surface spin
canting effect in IONPs

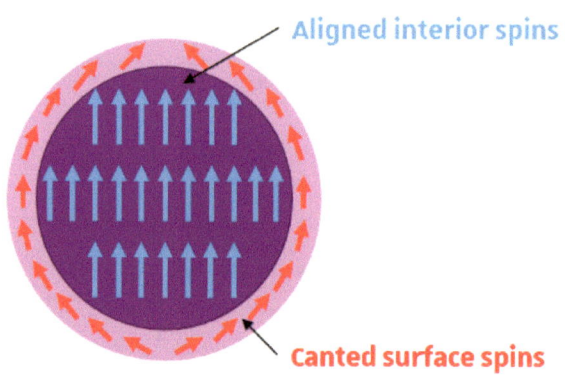

Aligned interior spins

Canted surface spins

and anisotropy of the NP and the measurement time depends on the technique and
apparatus used for measurement.

Effective Anisotropy Energy

A ferrimagnetic material can present magnetocrystalline anisotropy and shape
anisotropy. But going down to the nanoscale implies a new type of anisotropy that
was not present at the bulk scale: the surface anisotropy. This anisotropy is correlated
to the higher ratio between atoms on the surface and in the volume of the NP.

At the surface, atoms are sub-coordinated by comparison with core atoms and
due to the high surface curvature putting these atoms slightly out of equilibrium
position, spins of surface atoms tend to align their magnetic moments perpendicular
to the surface, which would lead to the spin canting effect (Fig. 13.6). Indeed, when
a magnetic field is applied, these spins do not align along the direction of the applied
magnetic field. The spin canting phenomenon is more important when IONPs get
smaller as the ratio between atoms located on the surface and those located in the
core is increased. This spin canting is one reason why the saturation magnetization
of IONPs is lower than that of the bulk phases. Defects and composition variation
are also responsible for lower M_S values observed with IONPs by comparison with
their bulk phases [14].

With superparamagnetic IONPs, an effective energy E_{eff} is often considered,
taking into account the magnetocrystalline anisotropy energy E_{MC}, the surface energy
E_s and a possible shape anisotropy energy E_{Sh}:

$$E_{eff} = E_{MC} + E_{Sh} + E_s \tag{13.5}$$

Therefore, the size, the shape and the composition influence greatly the overall
magnetic properties of IONPs. An increase in the NPs diameter ultimately leads to
an increase of the magnetocrystalline energy, as it is proportional to KV, with V the

volume of IONPs. The decrease of the surface energy with the size increase is generally overcome by the increase of energy due to the increase in NPs volume. It is also possible to change the effective anisotropy of a NP by modifying its shape. Therefore, the larger and more anisotropic the NPs are in shape, the higher the effective anisotropy energy is.

13.1.3   MRI with Iron Oxide Nanoparticles

IONPs have been commercially used as MRI T_2 contrast agents [73, 74] and are of particular interest as biodegradable and nontoxic nano-objects compared to other contrast agents' families [73, 75]. When formulating NPs suspensions for MRI, NPs must face different issues; they must be functionalized with a ligand ensuring their colloidal stability in solution, a controlled aggregation state, an optimal diffusion of water molecules near the magnetic core, a good biodistribution, and NPs must display a high saturation magnetization [76]. Hyeon's review [77] demonstrated that there is still enormous potential in the NPs-based MRI contrast agents. There is currently a real need to develop these products for early, accurate, rapid and targeted diagnosis of the suspected disease and MRI is nowadays among the most used imaging techniques in clinical diagnosis.

MRI is a non-invasive medical imaging technique that allows having a 2D or 3D view of a part of the body to get access to anatomic images of soft tissues. The major advantages of this method are its non-invasiveness and its limitless depth of exploration. However, acquisition time can be long and sensitivity is sometimes weak.

MRI is based on the phenomenon of nuclear magnetic resonance of hydrogen atoms. When submitted to an intense magnetic field, the overall magnetization is parallel to its direction. When a radiofrequency field is also applied perpendicular to the external one, then the overall magnetization has two directions: longitudinal magnetization M_z (parallel to the external magnetic field), and transverse magnetization M_{xy} (perpendicular to the magnetic field). Thus, when the radiofrequency field is applied, M_{xy} increases. When it stops, the spins return to their initial state: this is a relaxation phenomenon. From this can be defined two relaxation times: longitudinal relaxation time T_1 which corresponds to the time required for the longitudinal magnetization M_z to recover toward 63% of its initial value and the transverse relaxation time T_2, which is the time required for the transverse magnetization M_{xy} to decrease toward 37% of its initial value (Fig. 13.7). This T_2 called "true" comparatively to T_2^* which also take into account the inhomogeneity of the magnetic field created inside the MRI scanner. T_2^* is calculated from FID signal (Free Induction Decay) acquire by the machine during a classical MRI sequence [78].

Figure 13.8 explains the difference between T_2 and T_2^*. The diagram shows the process of transverse relaxation after a 90° radiofrequency pulse is applied at equilibrium. Initially the transverse magnetization (red arrow) has a maximum amplitude as the population of proton magnetic moments (spins) rotate in phase. The amplitude

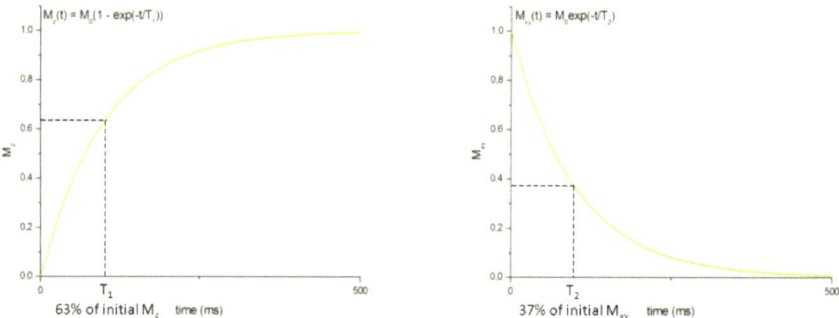

Fig. 13.7 Schematic representation of relaxation process with longitudinal relaxation at the left from which T_1 is the time required for the longitudinal magnetization to recover 63% of its initial value and with transverse relaxation at the right from which T_2 is the time required for the drop of 37% of transverse magnetization created by the frequency field

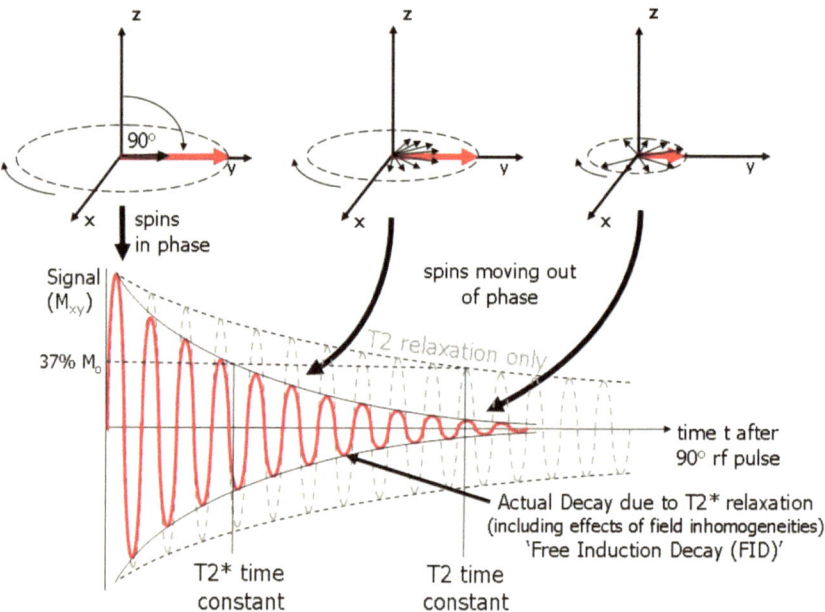

Fig. 13.8 Transverse (T_2 and T_2^*) relaxation process [79]. Reprinted with permission from JCMR

of the net transverse magnetization (and therefore the detected signal) decays as the proton magnetic moments move out of phase with one another (shown by the small black arrows). The resultant decaying signal is known as the Free Induction Decay (FID). The overall term for the observed loss of phase coherence (de-phasing) is T_2^* relaxation, which combines the effect of T_2 relaxation and additional de-phasing caused by local variations (inhomogeneities) in the applied magnetic field. T_2 relaxation is the result of spin–spin interactions and due to the random nature of molecular motion, this process is irreversible. T_2^* relaxation accounts for the more rapid decay of the FID signal, however the additional decay caused by field inhomogeneities can be reversed by the application of a 180° refocusing pulse [79]. Considering that T_2^* sequences are more impacted by inhomogeneities than T_2 ones; contrast agents, such as IONPs are used on T_2^*-weighted images.

Magnetic resonance (MR) images are constructed from these relaxation processes as approximately 65% of our body is composed of water. Contrasts are due to endogenous differences in water content in soft tissues, relaxation times, and diffusion characteristics of the analyzed tissues. The contrast can be improved by changing the parameters of the image acquisition method. However, sometimes contrast agents (CAs) are needed to increase sensitivity and help for the diagnosis. Their role is to shorten the relaxation time of surrounding hydrogen nuclei so that the contrast is increased between the areas influenced by the CA and those that are not. They give a better tissue characterization, they can reduce image artefacts or even give a functional information. Contrast agent shorten both T_1 and T_2, but because of their own relaxation rate which has an impact on the overall proton's relaxation rate, CAs affect more T_1 or T_2. This is why the use of the relaxivity r_2/r_1 ratio is important, with r_x is the inverse of T_x (s^{-1}). A good T_2 contrast agent have a greater effect on T_2 that on T_1, or a high r_2/r_1 ratio and induces a dark contrast. At the opposite, a T_1 CA have a low r_2/r_1 ratio and induce a white contrast. The diffusion of the water molecules toward the CA is also an important parameter for its effect on proton relaxation [80, 81]. In the context of molecular imaging, the specificity of MRI can be increased by directing CA to specific molecular entities.

High concentration of CA on site is needed because of the intrinsically low sensitivity of MRI (need high affinity, specificity and relaxivity). Paramagnetic Gd-complexes are T_1 CAs and IONPs have a higher effect on T_2 [82, 83]. Some known T_1 CAs are Gd complexes due to their paramagnetic properties. Gd complexes are often used in clinic but IONPs have a better biocompatibility and are less toxic [84]. Moreover, depending on their size, IONPs can be used both as T_1 and T_2 CAs. Indeed, even if the T_2 effect is still stronger than the T_1, r_2/r_1 ratio decreases with the IONPs diameter: ultrasmall IONPs (core of 4–6 nm) have a stronger effect on T_1 relaxation compared to bigger ones (even if they are still used as T_2 contrasting agent) [85, 86]. Compared to paramagnetic substances, the resultant magnetic moment of superparamagnetic particles is greater and responsible for a phenomenon of magnetic susceptibility disrupting the homogeneity of the external magnetic field. Negative CA are shortening T_2 (T_2^*) much more than T_1 of the nuclei situated in their neighborhood. T_2 will be reduced through the created field gradients and T_2^* effects will appear because of field inhomogeneities leading to a signal loss in the regions capturing the contrast agent on MR images.

Several types of negative CA accumulate in the liver when intravenously injected. After binding of plasma proteins (opsonins) to their surface, Endorem (Guerbet; 4.8–5.6 nm core and 80–150 nm hydrodynamic diameter) coated with dextran, and Resovist (SHU 555A, Bayer; 4.2 nm core, 62 nm hydrodynamic diameter) coated with carboxydextran, are captured by Kupffer cells within minutes. Since they are not retained in metastasis or hepatocytes, the darkening of liver signal (T_2) will only be observed in healthy parts of the organ, due to SPIO uptake by Kupffer cells. It is also possible to detect them in spleen and bone marrow macrophages. These IONPs were approved for detection of liver metastases in clinic before being taken off the EU and US market for economic reasons [87]. Ultrasmall particles of iron oxide (USPIO): Sinerem (Guerbet: 4–6 nm core + dextran coating = 20–40 nm hydrodynamic) has a lower r_2/r_1 ratio as compared to IONPs and are less likely to be captured by macrophages. They circulate longer, they can be used as blood pool agent for T_1. In late phase they accumulate in liver, spleen and become potent CA for lymphography (because of their small size they can cross the capillary wall and reach lymphatic system). Metastasis do not take up USPIO, they can be used to detect lymph nodes (Clariscan, GEHealthcare/Supravist, Bayer) [88]. It is worth noting that most of the commercial MRI contrast agent are multicore magnetic with significant influence on their superparamagnetic behaviour [89, 90]. In case of multi-core magnetic nanoparticles, the morphology of single core is less important, in comparison with magnetic properties. Therefore, the manifold analysis of the chosen IONPs synthesis is highly relevant [91].

13.2   Cancer Treatment with IONPs

Cancer is a global health problem that affect millions of people worldwide; with an increasing incidence. According to the Global Cancer Observatory (GLOBOCAN), cancer is the second leading cause of death globally, after ischemic heart disease; responsible for an estimated 9.6 million deaths in 2018. There are several types of cancer involving different mechanisms and with different properties or structures. Worldwide incidence, for both sexes and all ages, is given for 2020 in Fig. 13.9. Common factor between this heterogeneous large disease family is the abnormal cell grow presenting uncontrollably divisions leading to the formation of tumors.

The best way to treat cancer is to detect tumors precociously when their progression is not too important. This is possible with screening campaign, in breast cancer context for example and efficacy diagnostic tools. This early diagnosis allow physicians to offer rapid treatment without giving chances to the tumor to develop itself or metastasis.

Immune system's (IS) purpose is to protect an organism from diseases by distinguishing the organism's own healthy tissue from foreign organisms, such as bacteria or viruses, and unhealthy tissues such as tumoral cells. It is composed of two complementary components: the innate and the adaptative immune system [93]. Dendritic cells and phagocytes are rapidly recruited by innate IS to recognize a pathogenic agent, digest it and eliminate cellular debris. Adaptative IS is a slower mechanism but acts specifically against recognized pathogens via B and T lymphocytes. The antigenic escape is a mechanism that occurs when a host is unable to respond to

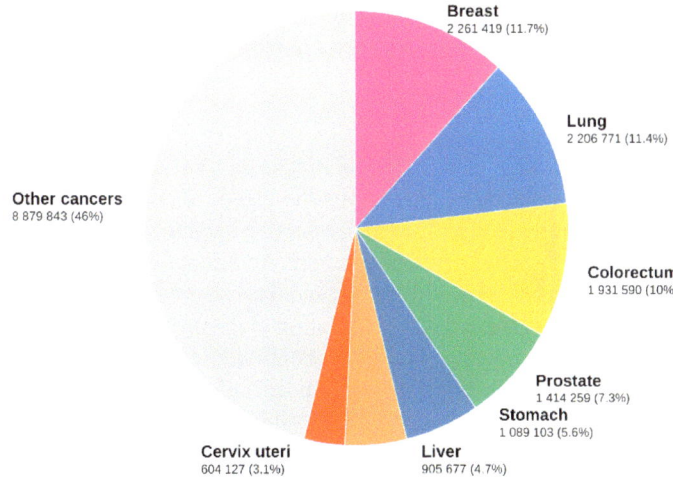

Total : 19 292 789

Fig. 13.9 Worldwide 2020 incidence for both sexes and all ages [92]. Reprinted with permission from WHO

an infectious agent. Immune evasion of tumoral cells involve the tumoral micro-environment (TME) which is highly studied in this beginning of twenty-first century, this field is known as immune-oncology.

Tumor's development depends on the composition of TME: endothelial cells (responsible of angiogenesis), immune cells (macrophages, lymphocytes, …), stromal cells (fibroblasts) and also acellular component such as cytokines or growth factors (i.e. epidermal growth factor). Epidermal growth factor (EGF) should be noted, its receptor (EGFR) is overexpressed in many type of cancers [94]. This receptor is involved in the tumoral proliferation induction and also in the antigenic escape of tumoral cells by inhibiting the process of presenting antigens of T-cells and down-regulating IS [95]. TME can have various compositions regarding the tumoral type or its degree of development. It is an indicator of patients prognostic [96].

Over the years, scientists and researchers have explored various treatment options for cancer. The chosen one will depend on the tumoral location, its stage, surgical accessibility, presence of metastasis and the patient's characteristics [97]. Usual methods are surgery, radiotherapy, chemotherapy or possibly immunotherapy and targeted therapy in a uni- or multi-modal way. In the cancer context, nanomedicines know an important development, in order to enhance contrast (for diagnosis) and to treat tumoral cells while conserving healthy tissue. Indeed, Gao et al. have shown an increase of papers published on the subject of nanomedicine in head and neck cancer (HNC) context; mostly with pre-clinical studies [98], but this it verifiable for each cancer's type.

13.2.1 Nanoparticles Designed for Theranostics: "We See What We Treat, and We Treat What We See" (Richard Baum)

Theranostic, the integration of therapy and diagnostics has emerged as a promising strategy in cancer management. It involves the use of diagnostic tools to identify tumor. This information can then be used to select a personalized treatment approach, such as targeted drug therapy or immunotherapy. Additionally, theranostic allows for real-time monitoring of treatment response and disease progression, enabling clinicians to adjust therapy as needed. The integration of therapy and diagnostics through theranostic represents a rising approach to improving cancer treatment outcomes while minimizing side effects. As research in this area continues to advance, it is hoped that theranostic will become an increasingly used tool in the fight against cancer [99, 100]. Nanoparticles (NPs) have emerged as promising candidates for cancer diagnosis and therapy due to their unique physicochemical properties and their ability to be functionalized with targeting moieties and therapeutic agents. NPs can be designed to specifically target cancer cells, deliver drugs and other therapeutic agents to the site of the tumor and enhance the therapeutic efficacy while minimizing toxicity to healthy tissues [101, 102]. NPs are classed according to their composition, size, what they carry on and their vectorization. It is possible to give 3 nanoparticle's families: lipidic-based, polymeric-based and inorganic-based NPs. Coating, shape, size, charge and composition can be different, making any classification difficult and not exhaustive [103] (Fig. 13.10). Lipidic nanoparticles is the category containing the highest number of FDA (Food and Drug Administration) approved nanoparticles with mostly liposome particles. Polymeric are rarely FDA approved but seems to be nice candidates for transporting small organic molecules, biological macromolecules and proteins or vaccines. They are soluble, biodegradable but can self-agglomerate and be toxic.

Lastly, inorganic nanoparticles made of gold (GNPs), iron oxide (IONPs) or silica have been studied for various drug delivery and imaging applications. There are wide variety of structures, geometries and shapes (nano-flowers,-cubes,-pellets,-sphere, ...). IONPs have the greatest number of FDA approved bio-applications in nanomedicine among inorganic nanoparticles. IONPs have unique magnetic properties that allow them to be detected by magnetic resonance imaging (MRI). Several studies have shown that IONPs can effectively detect cancer at an early stage, including HNC. For example, a study by Attia et al. demonstrated that magnetic nanoparticles could increase tumoral contrast (T_2^*-weighted MR images) on mice [104].

In order to offer a diagnosis or therapeutic ability, nanoparticles must reach and accumulate in tumoral area. After administration, they have to pass physical and biological barriers (i.e. epithelia, tumor micro-environment, IS) that limit the dispersal of nanoparticles. The nanoparticle's distribution will strongly impact their efficacy regardless the type of nanoparticles [103]. Their elimination is also an important issue, nanoparticles must not stay in the body for too long in order to

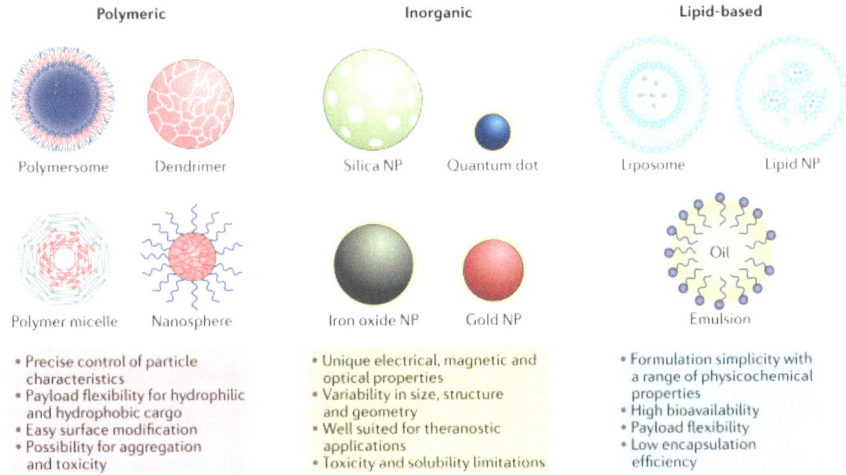

Fig. 13.10 Classes of nanoparticles [103]. Reprinted with permission from nature reviews, drug discovery

avoid any potential toxic effect. Generally, nanoparticles are quickly eliminated by the mononuclear phagocyte system (MPS) and monocytes. A poly-ethylene-glycol (PEG) coating can increase the nanoparticle's distribution time by escaping from MPS and avoiding aggregation issues [105]. It has been shown that the size and shape of nanoparticles impact their biodistribution. Smaller ones easily cross capillaries and are quickly eliminated by kidneys compared to larger ones (> 200 nm) that activate the complement system to be quickly removed from the blood stream and seem to induce more toxicity. Furthermore, nanoparticles < 20 nm have shown a higher tumoral penetration [106].

IONPs magnetic properties are known to be useful to enhance contrast on MRI, but they can also be used in therapy via magnetic hyperthermia (MH).

13.2.2 Magnetic Hyperthermia

Iron oxide nanoparticles are able to convert electromagnetic energy into heat when they are exposed to a high frequency alternating magnetic field. This released heat can kill, or weakens, cancer cells. Hyperthermia is defined by the transient increase in temperature above 37 °C. Depending on the temperature, two hyperthermic treatments have been developed: (i) mild hyperthermia, increasing temperature between 41° and 45 °C, preferentially induces the death of tumor cells, according to an apoptotic process, (ii) thermoablation, for a temperature above 45 °C, destroys tumors [107]. MH is also used to sensitize tumoral cells to another treatment, most of the time to chemotherapy [108, 109].

During the application of an alternating magnetic field, the nanoparticles are magnetized, their magnetic moments are progressively aligned in the direction of the magnetic field by rotation. When the magnetic field is reversed, the magnetic moments relax to their new position of equilibrium [110].

Some in vitro studies claim that the combination of magnetic nanoparticles with an alternating magnetic field (AMF) induces cell death [109, 111–117]. SAR values are observed with increasing frequencies and/or field amplitudes. Currently, most SAR values reported in the literature are measured at a frequency range of 300–700 kHz and a field range of 10–30 kA m^{-1}. However, for a safe application of hyperthermia to patients, a Brezovich criterion was at first established, where the product H*f should be less than 4.85×10^8 A m^{-1} s^{-1} to avoid eddy current effects (id. to avoid non-specific heating) [118]. By taking into account further technical improvements reducing eddy current heating, Hergt and Dutz [119] established another criterion H*f $= 5 \times 10^9$ A m^{-1} s^{-1} which is usually reported today. The experimental cancer models thus used have made it possible to demonstrate the concept of hyperthermia therapy using magnetic nanoparticles. Data from the literature show that it is possible to induce cell death without a detectable rise in temperature [111, 120, 121].

The challenge for in vivo applications is the possibility to eradicate cancer cells at the tumor site without damaging adjacent normal cells. Thus the pharmacokinetics, toxicity and biodistribution of nanoparticles are very important [122]. Several groups have studied magnetic hyperthermia in rodents and rabbits. A few human clinical trials have also been conducted [123–125]. NP administration consists either in injecting a NP suspension directly into the tumor, or intravenously, before applying the magnetic field. The validation of the concept of magnetic hyperthermia was first carried out by directly injecting the nanoparticles into the tumor. This approach allows to control the amounts of nanoparticles injected and to use relatively high doses compared to other modes of injection.

As we seen, in cancer context, one main challenge of theranostic nanomedicine is the specific targeting. The therapeutic distribution (drugs, heat, …) must be focalized in the tumoral neighboring in order to be fully efficient and to decrease the side-effect probability. Tumoral micro-environment must be taken into account because of the non-homogeneous and abnormal vasculature around tumors [126]. Two approaches exist in the delivery of nanoparticles: passive and active targeting.

13.3 Targeting Strategies

A major challenge in nanomedicine is to design NPs able to accumulate specifically in tumoral tissues without accumulating in clearance organs like the liver, the spleen, or the kidneys. There are two main mechanisms reported for the uptake of theranostic nanoplatforms in tumor sites. The first one is called "passive targeting", this targeting occurs because NPs, due to their size (diameter inferior to 100 nm) can pass easily through the abnormal vasculature of tumors. Indeed, they generally present irregular fenestrations and poor lymphatic drainage compared to healthy tissues. We talk about the enhanced permeability and retention (EPR) effect [104, 127]. However, the effectiveness of this effect depends on the tumor microenvironment but also on the capacity of NPs to accumulate there. Moreover, after intravenous (i.v.) injection, most developed NPs tend to accumulate in phagocytic organs such as the liver, spleen, and kidneys, and only small amounts are seen accumulated in tumors, as for example for iron oxide-based NPs [128–130]. Because of this limitation, a lot of research was done to develop "active targeting". The purpose is to selectively reach abnormal tissue, avoiding the uptake of NPs by healthy tissues. In this sense, the active targeting consists firstly in finding a receptor overexpressed in the type of cancer of interest. Once the target is defined, a specific molecule, known as the targeting ligand (TL), can be used to deliver the NPs directly to this site. These two modalities are resumed in the Fig. 13.11 [131–133].

It is worth nothing that the cellular internalization could be passive or active, depending on the presence of targeting ligands and can be modified by the shape or the size of nanoparticles. A passive diffusion through membrane is possible for smaller particles, but an active transport is mandatory for bigger ones.

13.3.1 Passive Targeting

EPR effect was discovered in 1980s by Maeda et al. It is the physiology-based main mechanism for large molecules and small particles to accumulate in tumors [127]. The blood vessels endothelium is fenestrated (200–800 nm) and becomes more permeable under certain conditions, such as inflammation or hypoxia, which is typical of tumor. Hypoxia promotes the formation of new blood vessels or the engulfment of existing ones by rapidly growing tumors. These newly formed vessels are leaky, allowing for selective enhanced permeation of macromolecules larger than 40 kDa and nanosystems to the tumor stroma. New vessels do not present a normal lymphatic drainage that contributes to the retention of NPs [134, 135]. Small low-molecular-weight agents, because of their ability to return to the circulation by diffusion, are not retained in tumors. Furthermore, the accumulation in tumoral site depends on nanoparticle's physico-chemical properties, such as size, surface charge and coating. The optimal NPs size range is around 20–200 nm to favor EPR effect. Particle diameter has a significant impact onto their biodistribution, as seen previously [136, 137]. Uptake by the MPS is a significant disadvantage in most therapeutic applications; a way to reduce the macrophage uptake is to coat nanoparticles with a hydrophilic polymer such as PEG which helps to reduce their opsonization by plasma protein [138]. We can cite Doxil© and Caelyx©; two pegylated liposomal drugs (doxorubicin) delivery system used in clinic that passively target breast and ovarian metastatic tumors.

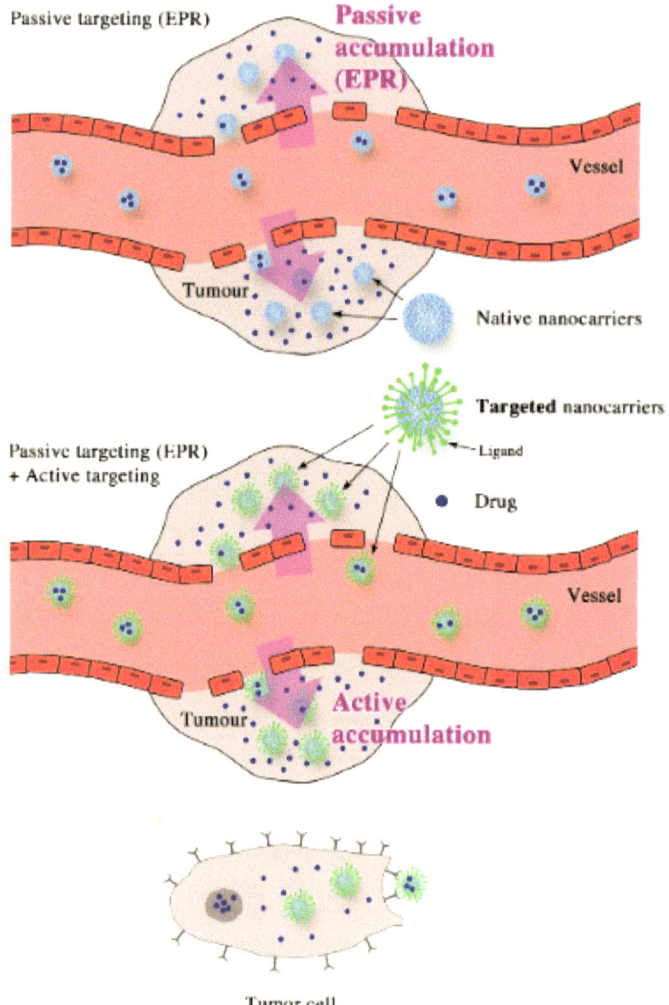

Fig. 13.11 Scheme illustrating the passive targeting by EPR effect and the active targeting into tumor site [104]. Reprinted with permission from Oxford University Press

13.3.2 Active Targeting

Among the most used receptors are the folate receptor, which can be targeted with folic acid on the surface of NPs [139] or the Human Epidermal Growth Factor Receptor 2 (HER2+) which is recognized by the monoclonal antibody Trastuzumab for example [140]. Peptides, nano-bodies, affibodies, proteins, small organic molecules, or nucleic acids constitute examples of short TL grafted to NPs [129, 131, 141–144]. The large antibodies or the smaller versions of them, the fragment antigen-binding (Fabs), are also common TL [145]. Iron oxide nanoparticles (IONPs), which accumulate in tumors without TL via intratumoral injection or passive targeting as a result of the EPR effect, are among the most developed and promising NPs at the moment [129–131, 133, 146–149]. When magnetic properties of the NPs are involved, another way to increase their accumulation is to apply a magnetic targeting, in which a magnetic field gradient guides IONPs toward the tumor; the main drawback of this technique is the requirement of a magnet that should be positioned easily near the tumor, thus limiting this strategy to specific cases. In this context, active targeting appears as a promising way for NPs selective internalization. However, Wilhem et al. observed that just 0.7% (median) of the i.v. administered nanoparticle dose (with or without TL) was found to be delivered to solid tumors after reviewing the literature from the previous 10 years [130]. Similar low percentages of internalized IONPs were also reported by Alphandery et al., who showed that many of them were accumulated in the liver and kidneys [146]. Therefore, there is a strong need to further understand how active targeting can be improved to increase the percentage of accumulated NPs in tumor sites after i.v. injection. This implies a better understanding of the key parameters that must be improved to raise the efficacy of IONPs bearing TL as active targeting systems. Among these parameters, the choice of the TL is essential. It is not an easy choice as it must target overexpressed receptors at the surface of cancer cells only. Moreover, the size of the TL must be suitable to the mean size of IONPs and the grafted TL must conserve its recognition capacity toward its target, a fact that imposes *sine qua non* conditions on the chosen bioconjugation approach to avoid the involvement of the active site in the coupling. In the case of short TLs, a major concern is to avoid abrupt changes in the 3D conformation of the molecule (upon conjugation to the NP) that may alter its interactions with the cellular receptors. Lastly, the TL must remain fully exposed to the solution and not buried inside the organic coating (such as PEG chains) already present on the IONPs surface. The coupling of the TL at the surface of IONPs is also an important step. There is a wide spectrum of bioconjugation reactions [150] used to attach TL at the surface of functionalized IONPs and; globally they can be divided into two types: covalent or non-covalent conjugations. The first group belongs to the carbodiimide chemistry [151], Michael addition [152], click Chemistry [153], or Diels–Alder cycloaddition [129, 154], among others. In non-covalent conjugation, the most common ones are electrostatic interactions, metal affinity coordination, or biotin-avidin interaction [155]. The choice of reactions mainly depends on the type of molecule used as a TL and its eventual functional group that may be used to perform a covalent conjugation technique, which brings more stability for the grafting than a non-covalent technique. Furthermore, it is also important to validate the presence of TL at the surface of IONPs and to quantify it [156].

Commonly, the presence of TL at the surface of IONPs may be checked by FTIR spectroscopy or thermogravimetric analysis (TGA) if the amount of TL is high enough, but most of the time it is determined by the observation of an increase in the hydrodynamic size or a change of zeta potential values. Hence, the most challenging step is the quantification of the amount of TL at the surface of IONPs [131, 133, 143, 157]. The active endocytosis pathway can pass through clathrin-

or caveolin- mediated but also endocytosis without mediation. Clathrin involves the formation of vesicles after ligand-receptor contact. Caveolin takes part in the protein, lipidic and fatty acid regulation [158]. Svitkova et al. [159] have shown that bovine serum albumin (BSA) coated nanoparticles are mostly internalized by clathrin mediated endocytosis but PEG coated ones use caveolin and lipid raft for their internalization. EGFR has been introduced earlier as a good receptor to target for cancer uses because of its overexpression in tumoral cells and its implication in angiogenesis and tumorigenesis process. Study have shown that small molecular weight peptide presents a good affinity to EGFR [160].

Magnetic drug targeting is another way to actively target tumors. Using super-paramagnetic properties of IONPs, nanoparticles are guided through the body under the action of a localized magnetic field to specifically deliver therapeutic effects. Magnetic nanoparticles with high saturation magnetization responds better to the magnetic field, allowing an higher tumoral accumulation. It will strongly depend on the size and shape of nanoparticles [6, 71]. Freeman et al. [161] where firsts to use magnets to move iron particles through vascular system. More recently, Shen et al. [162] has proved a higher IONPs retention in the rat's heart when placed under localized external magnetic field. This retention seems to be proportional to the magnetic field strength.

With this better understanding of the biodistribution and internalization pathway of nanoparticles from the administration to the target; some examples of interesting multimodal nanoparticles will be presented in the next part.

13.4 Multimodal Nanoparticles

Multimodal nanoparticles are nanoparticles that have physico-chemical properties allowing diagnosis and/or therapeutic effect, such particles can carry multiple types of payloads, such as drug, imaging agents, and targeting ligands in a single particle. By definition theranostic nanoparticles are multimodal. These nanoparticles can offer several advantages over traditional cancer therapies, including improved targeting of cancer cells, enhanced drug efficacy, delimitation of tumors and reduced side effects. Figure 13.12. presents the different applications for IONPs. In recent years, the development of multimodal nanoparticles has gained significant interest in the cancer research community for their therapeutic and diagnostic applications allowing for real-time monitoring or drug delivery and tumor response [163].

In this context, understanding the properties and behavior of these NPs in biological systems is a key point for their successful translation into the clinic. In this part, several examples of nanoparticles classified by therapeutic effect will be seen.

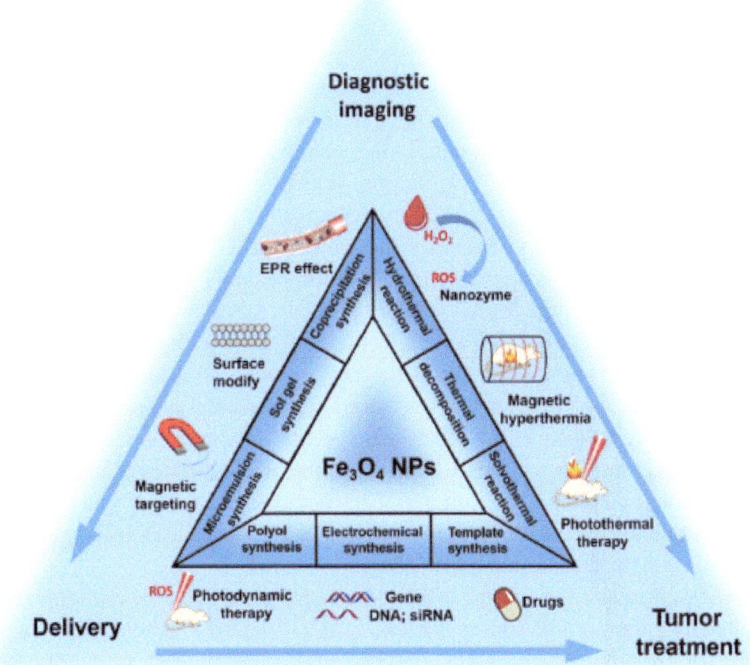

Fig. 13.12 IONPs applications in cancer diagnosis and treatment [163]. Reprinted with permission from Theranostics

13.4.1 Radiotheranostic

Radiation therapy (RT) is the main way to treat cancer, approximately 50% of all cancer patients receive radiation therapy [164]. This treatment uses high-energy radiation to kill cancer cells. It can be delivered by external beam (X-ray) or internal radiation therapy (brachytherapy) which involves placing a radioactive source inside the body, near the tumor. Theses radiations damage DNA and biological macromolecules, preventing cell proliferation and tumor growth. Healthy cells are impacted by these radiations but are, usually, more able to repair themselves and recover from the damages. Radiation therapy is often used in combination with surgery or chemotherapy to prevent resistance or tumoral recurrence. This is a non-specific treatment but always localized in the tumoral region. Despite the recent progress of radiation therapy with thinner ionizing beam, dose fragmentation, or Flash (ultra-high dose rate radiation); a lot of research are in course to reduce side-effects by decreasing the dose or by increasing the tumoral sensitivity to RT or to deliver in a very localized way radiation inside the tumor.

The aim of NPs combination with X-rays radiotherapy (XRT) or proton-therapy (PRT) is to potentialize irradiations effects and to increase specificity in order to decrease the distributed dose and spare surrounding healthy tissues. Many metallic nanoparticles have been studied for these application because of their high atomic number (Z) increasing the probability to interact with radiations [165, 166].

Hainfeld et al. in [167] were the firsts to develop gold nanoparticles to enhance radiotherapy in mice. They irradiated mice bearing subcutaneous mammary carcinomas with X-rays with and without pre-injection of GNPs. One-year survival was 86% for mice injected with GNPs versus 20% for the control group irradiated with the same energy. It shows the interest of GNPs combined with XRT. Furthermore, there was no apparent toxicity for mice, and GNPs were largely cleared from the body through the kidneys. More recently, Li et al. [168] have functionalized polyallylamine-coated GNPs with Cetuximab, a monoclonal antibody targeting and blocking epidermal growth factor receptor (EGFR), thus inhibiting the tumoral development. For the first time, an enhanced cellular uptake, thanks to Cetuximab combined with radio-enhancing effect was successfully tested in vitro. Figure 13.13 shows the direct damages to DNA by photon (red wiggly line) and ion (red straight line) radiations; but also mitochondria. The impacted medium produces secondary electrons, radicals and reactive species that will also impact DNA, organelles and macromolecules (indirect damages) [166].

Tumor radiosensitization was also studied on iron oxide nanoparticles (IONPs) irradiated by gamma-radiation by Shetake et al. [169]. After nanoparticles injection, the tumoral growth was significantly decreased. Furthermore, this kind of NPs have the advantage to induce a T_2^*-weighted MRI contrast which allows their monitoring inside the patient's body. Ahmad et al [170] have compared the radiosensitizing effect of three commercially available NPs (GNPs, IONPs and Gd-complex) on tumoral cells (MCF-7 and U87). After measuring the NPs uptake and cytotoxicity; they shown a higher enhancement factor (radiosensitizer effect) with GNPs on U87 and no difference on MCF-7. IONPs main advantage is to be monitored directly in the patient's body via MRI.

Ternad et al. [171] have studied IONPs radiosensitizing properties [171] on tumoral cells (A549) exposed to 225 kV X-rays. As results, authors describe that

Fig. 13.13 Illustration of radiation damage mechanisms [166]. Reprinted with permission from Cancer Nanotechnology

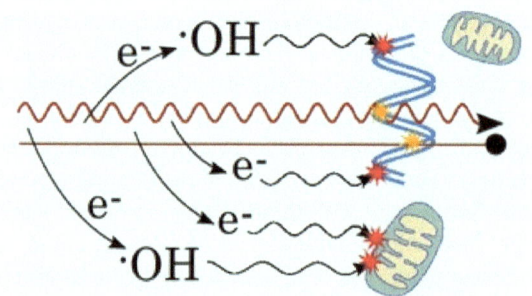

radiosensitization does not result only from a physical phenomenon but that an important part comes from biological events. They demonstrated an inhibition of thioredoxin reductase enzymes that prevent the regeneration of intracellular antioxidant and have a role in the management of oxidative stress. Inhibiting these proteins increase the effect of the radiation therapy by preventing cells from self-repairing.

13.4.2 Chemotherapeutic Nanoparticles

Chemotherapy is a main treatment option for cancer after surgery even if side-effects have been reported since the beginning of its use on tumors in the 1940s by Louis S. Goodman and Alfred Gilman. It is a treatment that uses chemicals to destroy cancer cells in the patient's body. Chemicals used have as main target cells that divide and grow rapidly; which is the case for tumoral cells. These drugs work by interfering with the cell division process and prevent it. Chemotherapy is usually given in cycles, with a period of treatment followed by a period of rest to allow the body to recover. Indeed, this is a non-specific, or a systemic, treatment that attacks every dividing cells, which causes many side-effects and decrease significantly the patient's quality of life. Chemotherapy is often combined with other treatment such as surgery or radiation therapy because of the potential chemo-resistance of tumoral cells, furthermore, several drugs are frequently administered extemporaneously with different modes of operation [172]. Depending on the location of the tumor, the combination of drugs is different. For example in lung cancer, it is recommended to deliver cyclophosphamide, doxorubicin, vincristine and prednisolone but for colorectal cancer, 5-fluorouracil, folic acid and oxaliplatin are prescribed [173]. Despite the efficacy of chemotherapy, it is not used for every patient because of chemo-resistances which can occur due to various factor such as genetic mutations, tumor-microenvironment, or the type of tumoral cells. Intrinsic resistance is the resistance due to the tumor type, for example triple negative breast cancer has a decreased responsiveness to drugs. We can also quote gastric cancer overexpressing HER2 that are known to be resistant to cisplatin [174]. Acquired resistance is a gradual reduction of anticancer efficacy of a drug during treatment. It may result of mutation due to the drug, a second proto-oncogene activation or changes in TME after treatment. As example, neuroblastoma cancer cells release exosomes to the TME that induce production of miRNAs by tumor-associated macrophages (TAMs) after cisplatin treatment. These miRNAs will silence the TERF1 gene of neuroblastoma cells, increasing their telomerase activity and their resistance to chemotherapy [175]. A scheme of some resistance mechanisms examples is given in Fig. 13.14 [176].

Therefore, there is a need to develop new strategies to deliver chemotherapeutic drugs in tumoral cells that are resistant or have developed a resistance and also to deliver them locally in order to reduce side-effect by decreasing the concentration in the body. In this context, nanoparticles seem to be good candidates.

Huang et al. [177] develop IONPs co-coated with PEG and PEI polymers design for dual target-specific drug delivery and MRI in cancer theranostic [177]. Targeting

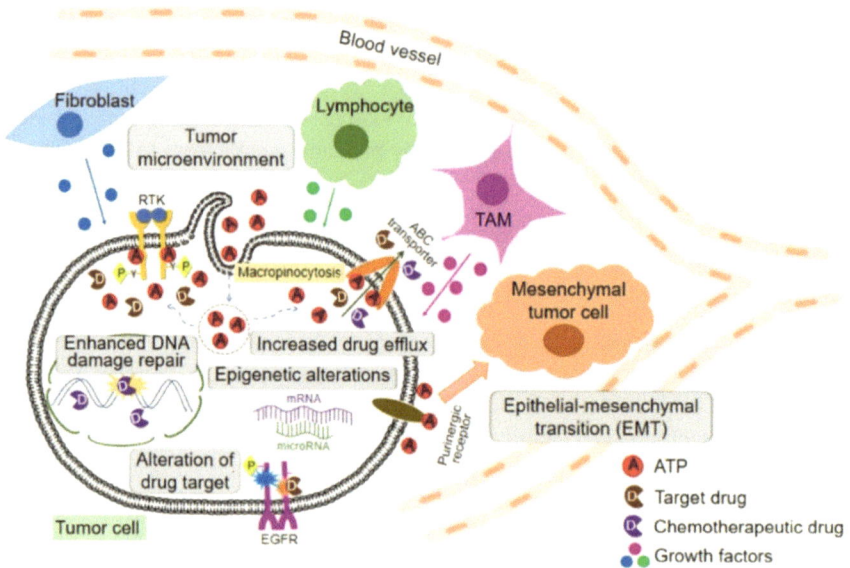

Fig. 13.14 Cancer cells mechanisms involved in drug resistance [176]. Reprinted with permission from cancer nanotechnology

part of these NPs is supported by conjugation of folic acid which receptor is over-expressed in various human carcinomas. IONPs are here loaded with doxorubicin and injected on MCF-7 tumoral xenografted mice. Authors describe a nice IONPs tumoral uptake but with a predictable accumulation in the liver. The efficacy of this drug-delivery system was proven by a stable tumoral growth 35 days post-injection for the group Dox-loaded IONPs and also by the increased presence of IONPs in tumors ex vivo for NPs conjugated with folic acid compared to NPs without targeting ligand. Smart nanocarrier for multi-stimuli on-demand drug delivery have also been synthesize by Elsami et al. [178]. Flower-like IONPs were encapsulated in a dual pH and thermoresponsive responsive copolymer to release drugs (Dox.) in a highly controlled way. A negligible amount of doxorubicin was released from nanoparticles in physiological conditions (37 °C and neutral pH). But increasing the temperature and decreasing pH by magnetic hyperthermia, Dox. was massively released from NPs in their surrounding environment; suggesting that IONPs can be used to remote-control drug release in combination with MH for cancer treatment.

El-Dakdouki's team develop hyaluronan-coated iron oxide nanoparticles (HA-IONPs) loaded with doxorubicin for preclinical. Hyaluronan was chosen to target CD44, a cell surface glycoprotein expressed on many cancer cells. An accumulation of NPs; darkening inside the yellow circle is observed on tumoral mice (Fig. 13.15) 1 h post-injection and still observable 24 h after. The injection of Dox-HA-IONP inhibits the tumoral growth approximately 4 times more than free doxorubicin. Efficacy is proved to be higher than standard Dox even with an inferior doxorubicin

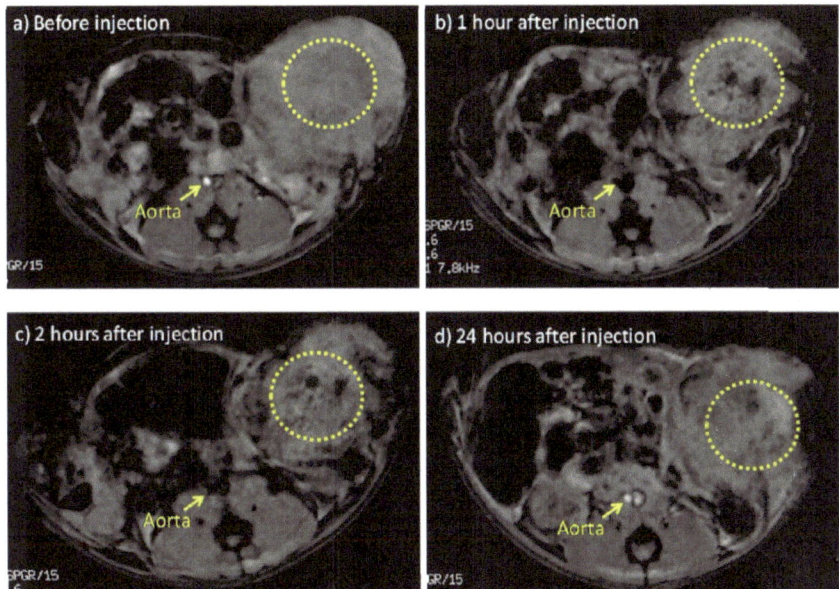

Fig. 13.15 T_2^*-weighted MR images of mouse tumor before injection, 1, 2 and 24 h after HA-IONP injection. Negative contrast is highlighted in yellow circle suggesting the presence of IONPs [180]. Reprinted with permission from ACS

concentration (2 mg_{Dox}/kg_{body} for condition Dox-HA-IONP versus 8 mg_{Dox}/kg_{body} for the condition free Dox.) implying less side-effects [179, 180].

Some nanoparticles are tested in clinical study as drug delivery system; because of their biocompatibility, lipidic-based NPs are the firsts to reach this step. These studies will serve as proof-of-concept for IONPs drug delivery system. We can cite the phase II clinical study, on 98 Korean women presenting ovarian cancer, Lee et al. [181] have used cremophor-free polymeric micelle with paclitaxel as first-line treatment [181]. They compared the efficacy of their micellar formulations of paclitaxel to a generic treatment with paclitaxel. As conclusion, they showed a non-inferior efficacy and less side-effects because of the presence of micelles enhancing solubility of paclitaxel and allowing a higher accumulation of drugs on tumoral site.

13.4.3   Magnetic Hyperthermia Activated Drug Delivery

As shown previously in this chapter, magnetic nanoparticles enable specific response when submitted to high frequency (magnetic hyperthermia) or low frequency (magneto-mechanical therapy) alternating magnetic field (AMF). As other metallic nanoparticles, they produce reactive oxygen species, intrinsically or under ionizing beam, making them interesting for cancer treatment.

Guisasola et al. [182] used IONPs embedded in a mesoporous silica matrix and coated with an engineered thermoresponsive polymer. Under AMF, temperature increase in the NPs neighboring causing the polymer transition and the consequent release of drug (Doxorubicin) trapped inside the silica pores. The therapeutic efficacy is not based on the tumoral tissue heating which avoids the necessity to employ large amount of magnetic cores as is common in current MH. Furthermore, the chemotherapeutic agent is delivered in the tumoral region decreasing the side-effect risk [182]. The tumoral growth monitored is significantly decrease for the condition: magnetic hyperthermia + doxorubicin compared to conditions MH or Dox. alone. The combination of both treatments (chemotherapy and MH) confers a higher efficacy and less side-effects.

In 2021, Fang et al. [183] have developed magnetic liposomal systems conjugated with a targeting ligand and loaded with a immunotherapeutic drug (CSF1R inhibitor). The combination of MH and M2 macrophage repolarization in tumoral microenvironment relieves tumoral immunosuppression, normalizes tumor blood vessels and promotes the infiltration of T-lymphocytes. After the treatment, an increase of antitumoral effector $CD8^+$ T cells was also observed. Thus, TME was remodeled, nanoparticles have also activated immune response and memory inhibiting tumoral recurrence.

13.4.4 Photoresponsive Nanoparticles

Development of photosensitizers multimodal NPs in the cancer therapy context is a topic in expansion. Upon UV–vis or near-infrared (NIR) light, metallic nanoparticles can heats their surrounding environment [184]. Figure 13.16a illustrates the production of reactive oxygen species (ROS) due to photochemical reactions with oxygen after exposure to UV light, this mechanism is known as photodynamic therapy (PDT). On Fig. 13.16b, the photothermal therapy (PTT) is represented; after irradiation with NIR light, nanoparticles heat surrounding cells. Both techniques are minimally invasive but have a major limitation, the low light penetration through tissues [185]. The first team that has shown the conversion of light absorbed by IONPs to local heating was Yu et al. [186]. Under NIR irradiation, alumina-coated iron oxide magnetic nanoparticles were used as photothermal agents to selectively kill bacteria. After 5 min of light irradiation, the temperature increased by 20 °C and decreased nosocomial bacteria growth (Gram positive and negative and antibiotic-resistant) by over 95% within 10 min of light irradiation. The possibility to do PTT with nanoparticles was also proved on highly crystallized iron oxide nanoparticles coated with polysiloxane-containing copolymer [187] offering great antibiofouling properties and an enhanced tumoral accumulation through EPR effect. Different composition were successfully tested to induce PTT, such as carboxymethyl chitosan-coated or PEGylated IONPs; or plasmonic MXene-based nanocomposites [187–191].

Besides the possibility to heat or produce ROS after illumination, nanoparticles able to release drugs upon exposure to light have been developed. Wu et al.

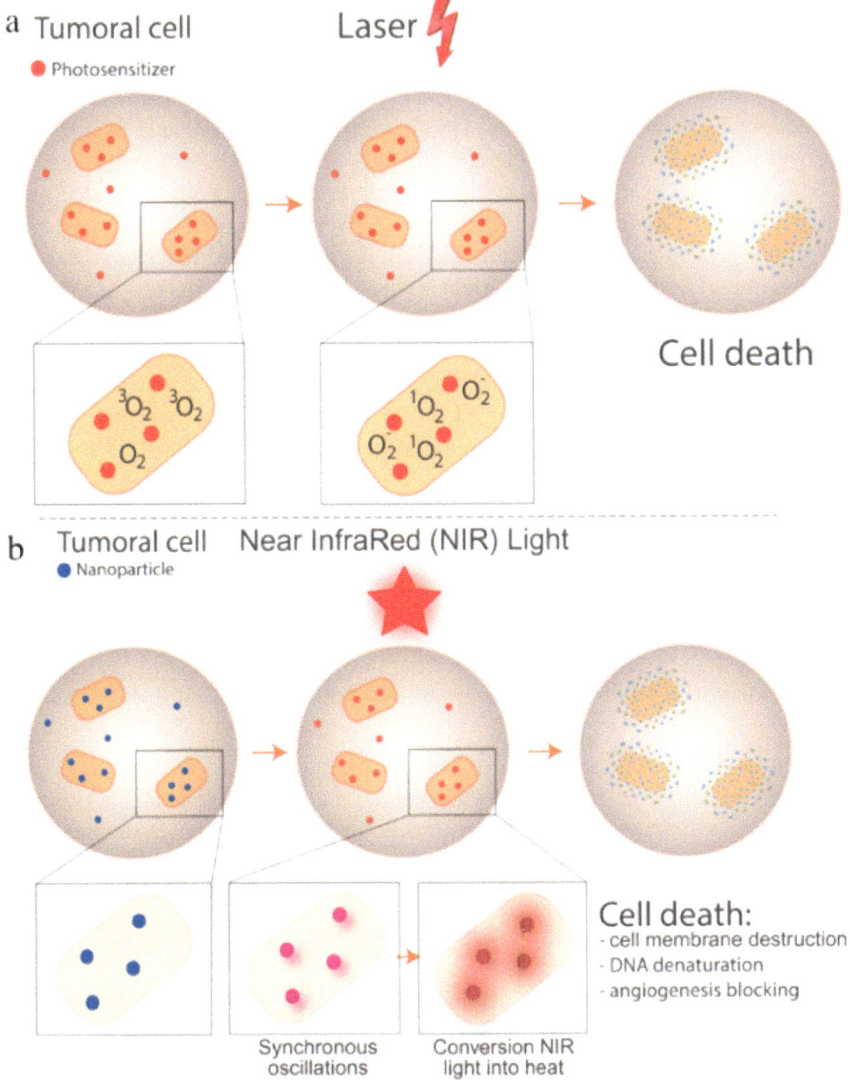

Fig. 13.16 Illustration of **a** photodynamic therapy and **b** photothermal therapy mechanisms [192]. Reprinted with permission from De Gruyter

[193] have loaded IONPs with doxorubicin via thermos-sensitive bond, cleaved by a near infrared exposure [193]. Using magnetic targeting, NPs are accumulated inside tumors; photothermal therapy is activated upon NIR exposure and doxorubicin is released. These nanoplatforms were tested on MCF-7 and on xenograft tumor Balb/c nude mice. Tumoral growth was significantly decrease using the combination of PTT and doxorubicin delivery compared to other groups. Sun et al. [194] have

also demonstrated the interest of using IONPs as chemo-photodynamic combination therapy [194]. Authors used an aptamer-hybridized nucleic acid structure to target tumoral cells; and IONPs were loaded with a chemical anticancer drug: daunomycin and a photosensitizer molecule. Upon visible light, PDT mechanism induces ROS production and daunomycin is released the tumoral micro-environment. They showed in vitro an accurate tumor targeting and high cytotoxicity after light exposure.

NIR illumination can also serve for real-time imaging. Kanwar et al. (2016) used iron oxide saturated lactoferrin nanocapsules (FebLfNCs) for real-time imaging and anti-tumoral therapy [195]. Iron saturated form of lactoferrin have outstanding immune-modulatory properties: interleukins productions, activation of macrophages, natural killer (NK) cells, dendritic cells. FebLFNCs are emit fluorescence (675–740 nm) after NIR light exposure. Authors conclude in an efficacy drug-delivery system that can be monitored in vivo on tumoral mice model with NIR fluorescence imaging. It could be interested to implement PTT or PDT to this kind of theranostic nanoplatforms in order to increase its therapeutic ability. Many preclinical studies can be found on the use of nanoparticles in photothermal therapy, it is possible to cite Wu et al. working on the combination of hyperthermia and drug-release with pegylated silica-core gold nanoshells exposed to an external NIR laser irradiation [196]. But few examples have reached clinical trial.

13.5 Conclusion

The synthesis and characterization of iron oxide nanoparticles (IONPs) have been the focus of intense research due to their unique magnetic properties and potential applications in a wide range of fields, including biomedical, environmental, and industrial areas.

On this chapter, the synthesis and characterization of IONPs have been studied. The ability to control their size and shape using various synthesis methods has been a key aspect of their development, as it allows for the fine-tuning of their physico-chemical properties. One of the main applications of IONPs is in magnetic resonance imaging (MRI) where these properties can be exploited to enhance contrast and improve the sensitivity and specificity of this imaging tool. IONPs also shown a great potential as theranostic nanoplatforms. Magnetic hyperthermia (MH) uses IONPs to selectively heat tumor tissue, leading to the destruction (or weakening) of tumoral cells. An explanation of this therapy was given here and various parameters have been identified to optimize IONPs magnetic properties for improved cancer diagnosis and therapy.

An overview of the cancer problematic was given; introducing the main challenges of tumoral treatments (surgery, chemotherapy and radiation therapy). Main issue of nanomedicine is to target tumoral cells. An overview of passive and active targeting was given here, with a highlight for magnetic drug targeting. Specific targeting is a key point to increase anti-tumoral efficacy and decrease side-effects. Currently, various IONPs nanoplatforms are developed notably for diagnostic: MRI, radiolabeled for PET/SPECT (Positron emission tomography and Single-photon emission computed tomography) or real-time monitoring under near-infrared exposure; and therapeutic: MH, photodynamic (PDT) and photothermal therapy (PTT) or as drug delivery system.

References

1. S.M. Dadfar, K. Roemhild, N.I. Drude, S. Stillfried, R. Knüchel, F. Kiessling, T. Lammers, Adv. Drug Deliv. Rev. **138**, 302–325 (2019)
2. A. Ali, H. Zafar, M. Zia, I. ul Haq, A.R. Phull, J.S. Ali, A. Hussain, Nanotechnol. Sci. Appl. **9**, 49–67 (2016)
3. X.L. Dong, C.J. Choi, B.K. Kim, J. Appl. Phys. **92**, 5380–5385 (2002)
4. R. Massart, IEEE Trans. Magn. **17**, 1247–1248 (1981)
5. W. Ling, M. Wang, C. Xiong, D. Xie, Q. Chen, X. Chu, X. Qiu, Y. Li, X. Xiao, J. Mater. Res. **34**, 1828–1844 (2019)
6. D. Stanicki, T. Vangijzegem, I. Ternad, S. Laurent, Exp. Opin. Drug Deliv. **19**, 321–335 (2022)
7. T. Hyeon, S.S. Lee, J. Park, Y. Chung, H.B. Na, J. Am. Chem. Soc. **123**, 12798–12801 (2001)
8. S. Sun, H. Zeng, J. Am. Chem. Soc. **124**, 8204–8205 (2002)
9. C.B. Murray, D.J. Norris, M.G. Bawendi, J. Am. Chem. Soc. **115**, 8706–8715 (1993)
10. X. Peng, J. Wickham, A.P. Alivisatos, J. Am. Chem. Soc. **120**, 5343–5344 (1998)
11. V.K. LaMer, R.H. Dinegar, J. Am. Chem. Soc. **72**, 4847–4854 (1950)
12. Y. Xia, Y. Xiong, B. Lim, S.E. Skrabalak, Angew. Chem. Int. Ed. Engl. **48**, 60–103 (2009)
13. S.G. Kwon, Y. Piao, J. Park, S. Angappane, Y. Jo, N.-M. Hwang, J.-G. Park, T. Hyeon, J. Am. Chem. Soc. **129**, 12571–12584 (2007)
14. W. Baaziz, B.P. Pichon, S. Fleutot, Y. Liu, C. Lefevre, J.-M. Greneche, M. Toumi, T. Mhiri, S. Begin-Colin, J. Phys. Chem. C **118**, 3795–3810 (2014)
15. Y. Xia, Y. Xiong, B. Lim, S.E. Skrabalak, Angewandte Chemie (International ed. in English) **48**, 60–103 (2009)
16. F. Perton, G. Cotin, C. Kiefer, J.-M. Strub, S. Cianferani, J.-M. Greneche, N. Parizel, B. Heinrich, B. Pichon, D. Mertz, S. Begin-Colin, Inorg. Chem. **60**, 12445–12456 (2021)
17. H. Chang, B.H. Kim, H.Y. Jeong, J.H. Moon, M. Park, K. Shin, S.I. Chae, J. Lee, T. Kang, B.K. Choi, J. Yang, M.S. Bootharaju, H. Song, S.H. An, K.M. Park, J.Y. Oh, H. Lee, M.S. Kim, J. Park, T. Hyeon, J. Am. Chem. Soc. **141**, 7037–7045 (2019)
18. G. Cotin, B. Heinrich, F. Perton, C. Kiefer, G. Francius, D. Mertz, B. Freis, B. Pichon, J. Strub, S. Cianférani, N. Ortiz Peña, D. Ihiawakrim, D. Portehault, O. Ersen, A. Khammari, M. Picher, F. Banhart, C. Sanchez, S. Begin-Colin, Small 2200414 (2022)
19. J. Park, K. An, Y. Hwang, J.-G. Park, H.-J. Noh, J.-Y. Kim, J.-H. Park, N.-M. Hwang, T. Hyeon, Nat. Mater. **3**, 891–895 (2004)
20. S. Sun, H. Zeng, D.B. Robinson, S. Raoux, P.M. Rice, S.X. Wang, G. Li, J. Am. Chem. Soc. **126**, 273–279 (2004)
21. F.X. Redl, C.T. Black, G.C. Papaefthymiou, R.L. Sandstrom, M. Yin, H. Zeng, C.B. Murray, S.P. O'Brien, J. Am. Chem. Soc. **126**, 14583–14599 (2004)
22. A. Demortière, P. Panissod, B.P. Pichon, G. Pourroy, D. Guillon, B. Donnio, S. Bégin-Colin, Nanoscale **3**, 225–232 (2011)
23. E. Wetterskog, C.-W. Tai, J. Grins, L. Bergström, G. Salazar-Alvarez, ACS Nano **7**, 7132–7144 (2013)
24. A. Lak, M. Kraken, F. Ludwig, A. Kornowski, D. Eberbeck, S. Sievers, F.J. Litterst, H. Weller, M. Schilling, Nanoscale **5**, 12286 (2013)

25. B.P. Pichon, O. Gerber, C. Lefevre, I. Florea, S. Fleutot, W. Baaziz, M. Pauly, M. Ohlmann, C. Ulhaq, O. Ersen, V. Pierron-Bohnes, P. Panissod, M. Drillon, S. Begin-Colin, Chem. Mater. **23**, 2886–2900 (2011)

26. H.T. Hai, H.T. Yang, H. Kura, D. Hasegawa, Y. Ogata, M. Takahashi, T. Ogawa, J. Colloid Interface Sci. **346**, 37–42 (2010)

27. X. Sun, N. Frey Huls, A. Sigdel, S. Sun, Nano Lett. **12**, 246–251 (2012)

28. A. Walter, C. Billotey, A. Garofalo, C. Ulhaq-Bouillet, C. Lefèvre, J. Taleb, S. Laurent, L. Vander Elst, R.N. Muller, L. Lartigue, F. Gazeau, D. Felder-Flesch, S. Begin-Colin, Chem. Mater. **26**, 5252–5264 (2014)

29. W. Baaziz, B.P. Pichon, Y. Liu, J.-M. Grenèche, C. Ulhaq-Bouillet, E. Terrier, N. Bergeard, V. Halté, C. Boeglin, F. Choueikani, M. Toumi, T. Mhiri, S. Begin-Colin, Chem. Mater. **26**, 5063–5073 (2014)

30. W. Baaziz, B.P. Pichon, C. Lefevre, C. Ulhaq-Bouillet, J.-M. Greneche, M. Toumi, T. Mhiri, S. Bégin-Colin, J. Phys. Chem. C **117**, 11436–11443 (2013)

31. G. Cotin, C. Kiefer, F. Perton, M. Boero, B. Özdamar, A. Bouzid, G. Ori, C. Massobrio, D. Begin, B. Pichon, D. Mertz, S. Begin-Colin, ACS Appl. Nano Mater. **1**, 4306–4316 (2018)

32. A.N. Shipway, E. Katz, I. Willner, ChemPhysChem **1**, 18–52 (2000)

33. D.A.J. Herman, S. Cheong-Tilley, A.J. McGrath, B.F.P. McVey, M. Lein, R.D. Tilley, Nanoscale **7**, 5951–5954 (2015)

34. G. Cotin, F. Perton, C. Petit, S. Sall, C. Kiefer, V. Begin, B. Pichon, C. Lefevre, D. Mertz, J.-M. Greneche, S. Begin-Colin, Chem. Mater. **32**, 9245–9259 (2020)

35. M.V. Kovalenko, M.I. Bodnarchuk, R.T. Lechner, G. Hesser, F. Schäffler, W. Heiss, J. Am. Chem. Soc. **129**, 6352–6353 (2007)

36. U. Jeong, X. Teng, Y. Wang, H. Yang, Y. Xia, Adv. Mater. **19**, 33–60 (2007)

37. J. van Embden, A.S.R. Chesman, J.J. Jasieniak, Chem. Mater. **27**, 2246–2285 (2015)

38. S.G. Kwon, T. Hyeon, Small **7**, 2685–2702 (2011)

39. G. Cotin, C. Kiefer, F. Perton, D. Ihiawakrim, C. Blanco-Andujar, S. Moldovan, C. Lefevre, O. Ersen, B. Pichon, D. Mertz, S. Bégin-Colin, Nanomaterials **8**, 881 (2018)

40. L.M. Bronstein, J.E. Atkinson, A.G. Malyutin, F. Kidwai, B.D. Stein, D.G. Morgan, J.M. Perry, J.A. Karty, Langmuir **27**, 3044–3050 (2011)

41. G. Salas, C. Casado, F.J. Teran, R. Miranda, C.J. Serna, M.P. Morales, J. Mater. Chem. **22**, 21065–21075 (2012)

42. P. Guardia, R. Di Corato, L. Lartigue, C. Wilhelm, A. Espinosa, M. Garcia-Hernandez, F. Gazeau, L. Manna, T. Pellegrino, ACS Nano **6**, 3080–3091 (2012)

43. P. Guardia, A. Riedinger, S. Nitti, G. Pugliese, S. Marras, A. Genovese, M.E. Materia, C. Lefevre, L. Manna, T. Pellegrino, J. Mater. Chem. B **2**, 4426–4434 (2014)

44. G. Cotin, C. Blanco-Andujar, F. Perton, L. Asín, J.M. de la Fuente, W. Reichardt, D. Schaffner, D.-V. Ngyen, D. Mertz, C. Kiefer, F. Meyer, S. Spassov, O. Ersen, M. Chatzidakis, G.A. Botton, C. Hénoumont, S. Laurent, J.-M. Greneche, F.J. Teran, D. Ortega, D. Felder-Flesch, S. Begin-Colin, Nanoscale **13**, 14552–14571 (2021)

45. I. Castellanos-Rubio, O. Arriortua, D. Iglesias-Rojas, A. Barón, I. Rodrigo, L. Marcano, J.S. Garitaonandia, I. Orue, M.L. Fdez-Gubieda, M. Insausti, Chem. Mater. **33**, 8693–8704 (2021)

46. R. Chen, M.G. Christiansen, A. Sourakov, A. Mohr, Y. Matsumoto, S. Okada, A. Jasanoff, P. Anikeeva, Nano Lett. **16**, 1345–1351 (2016)

47. L. Qiao, Z. Fu, J. Li, J. Ghosen, M. Zeng, J. Stebbins, P.N. Prasad, M.T. Swihart, ACS Nano **11**, 6370–6381 (2017)

48. L.M. Bronstein, X. Huang, J. Retrum, A. Schmucker, M. Pink, B.D. Stein, B. Dragnea, Chem. Mater. **19**, 3624–3632 (2007)

49. C.J. Meledandri, J.K. Stolarczyk, S. Ghosh, D.F. Brougham, Langmuir **24**, 14159–14165 (2008)

50. N.R. Jana, Y. Chen, X. Peng, Chem. Mater. **16**, 3931–3935 (2004)

51. J.M. Vargas, R.D. Zysler, Nanotechnology **16**, 1474–1476 (2005)

52. R. Hufschmid, H. Arami, R.M. Ferguson, M. Gonzales, E. Teeman, L.N. Brush, N.D. Browning, K.M. Krishnan, Nanoscale **7**, 11142–11154 (2015)

53. K.M. Kirkpatrick, B.H. Zhou, P.C. Bunting, J.D. Rinehart, Chem. Mater. **34**, 8043–8053 (2022)
54. T. Balakrishnan, M.-J. Lee, J. Dey, S.-M. Choi, CrystEngComm **21**, 4063–4071 (2019)
55. G. Cotin, C. Kiefer, F. Perton, M. Boero, B. Özdamar, A. Bouzid, G. Ori, C. Massobrio, D. Begin, B. Pichon, D. Mertz, S. Begin-Colin, A.C.S. Appl, Nano Mater. **1**, 4306–4316 (2018)
56. T. Belin, N. Guigue-Millot, T. Caillot, D. Aymes, J.C. Niepce, J. Solid State Chem. **163**, 459–465 (2002)
57. T.J. Daou, G. Pourroy, S. Bégin-Colin, J.M. Grenèche, C. Ulhaq-Bouillet, P. Legaré, P. Bernhardt, C. Leuvrey, G. Rogez, Chem. Mater. **18**, 4399–4404 (2006)
58. J.-P. Jolivet, E. Tronc, J. Colloid Interface Sci. **125**, 688–701 (1988)
59. N. Guigue-Millot, Y. Champion, M.J. Hÿtch, F. Bernard, S. Bégin-Colin, P. Perriat, J. Phys. Chem. B **105**, 7125–7132 (2001)
60. J. Santoyo Salazar, L. Perez, O. de Abril, L. Truong Phuoc, D. Ihiawakrim, M. Vazquez, J.-M. Greneche, S. Begin-Colin, G. Pourroy, Chem. Mater. **23**, 1379–1386 (2011)
61. M. Jeon, M.V. Halbert, Z.R. Stephen, M. Zhang, Adv. Mater. **33**, 1906539 (2021)
62. C. Hofmann, I. Rusakova, T. Ould-Ely, D. Prieto-Centurión, K.B. Hartman, A.T. Kelly, A. Lüttge, K.H. Whitmire, Adv. Funct. Mater. **18**, 1661–1667 (2008)
63. A. Lak, M. Cassani, B.T. Mai, N. Winckelmans, D. Cabrera, E. Sadrollahi, S. Marras, H. Remmer, S. Fiorito, L. Cremades-Jimeno, F.J. Litterst, F. Ludwig, L. Manna, F.J. Teran, S. Bals, T. Pellegrino, Nano Lett. 11 (2018)
64. A. Lak, M. Cassani, B.T. Mai, N. Winckelmans, D. Cabrera, E. Sadrollahi, S. Marras, H. Remmer, S. Fiorito, L. Cremades-Jimeno, Nano Lett. **18**, 6856–6866 (2018)
65. A. Lappas, G. Antonaropoulos, K. Brintakis, M. Vasilakaki, K.N. Trohidou, V. Iannotti, G. Ausanio, A. Kostopoulou, M. Abeykoon, I.K. Robinson, Phys. Rev. X **9**, 041044 (2019)
66. P. Guardia, A. Labarta, X. Batlle, J. Phys. Chem. C **115**, 390–396 (2011)
67. I. Castellanos-Rubio, I. Rodrigo, R. Munshi, O. Arriortua, J.S. Garitaonandia, A. Martinez-Amesti, F. Plazaola, I. Orue, A. Pralle, M. Insausti, Nanoscale **11**, 16635–16649 (2019)
68. R.M. Cornell, U. Schwertmann, *The Iron Oxides : Structure, Properties, Reactions, Occurrences, and Uses*, Second, Completely and Extended Edition (Wiley-VCH, Weinheim, 2003)
69. N.T.K. Thanh (ed.), *Clinical Applications of Magnetic Nanoparticles: Design to Diagnosis Manufacturing to Medicine* (CRC Press, Taylor & Francis Group, Boca Raton, 2018)
70. A.U. Gehring, H. Fischer, M. Louvel, K. Kunze, P.G. Weidler, Geophys. J. Int. **179**, 1361–1371 (2009)
71. J. Estelrich, E. Escribano, J. Queralt, M. Busquets, Int. J. Mol. Sci. **16**, 8070–8101 (2015)
72. N. Lee, T. Hyeon, Chem. Soc. Rev. **41**, 2575–2589 (2012)
73. S. Laurent, D. Forge, M. Port, A. Roch, C. Robic, L. Vander Elst, R.N. Muller, Chem. Rev. **108**, 2064–2110 (2008)
74. H.B. Na, I.C. Song, T. Hyeon, Adv. Mater. **21**, 2133–2148 (2009)
75. A.K. Gupta, M. Gupta, Biomaterials **26**, 3995–4021 (2005)
76. C. Blanco-Andujar, A. Walter, G. Cotin, C. Bordeianu, D. Mertz, D. Felder-Flesch, S. Begin-Colin, Nanomedicine (London, England), **11**, 1889–1910 (2016)
77. D. Kim, J. Kim, Y.I. Park, N. Lee, T. Hyeon, ACS Cent. Sci. **4**, 324–336 (2018)
78. W.A. Gibby, Neurosurg. Clin. N. Am. **16**, 1–64 (2005)
79. J.P. Ridgway, J. Cardiovasc. Magn. Reson. **12**, 71 (2010)
80. Z. Jászberényi, A. Sour, É. Tóth, M. Benmelouka, A. Merbach, Dalton Trans. 2713 (2005)
81. H. Duan, M. Kuang, X. Wang, Y.A. Wang, H. Mao, S. Nie, J. Phys. Chem. C **112**, 8127–8131 (2008)
82. P.A. Rinck, Mag. Res. Med
83. Y.-D. Xiao, R. Paudel, J. Liu, C. Ma, Z.-S. Zhang, S.-K. Zhou, Int. J. Mol. Med. **38**, 1319–1326 (2016)
84. J.W.M. Bulte, D.L. Kraitchman, NMR Biomed. **17**, 484–499 (2004)
85. Barbara Freis, G. Cotin, F. Perton, D. Mertz, S. Boutry, S. Laurent, S. Begin-Colin, Mag. Nanopartic. Human Health Med. 380–429 (2021)

86. L. Li, W. Jiang, K. Luo, H. Song, F. Lan, Y. Wu, Z. Gu, Theranostics **3**, 595–615 (2013)
87. J.W.M. Bulte, Am. J. Roentgenol. **193**, 314–325 (2009)
88. M. Kresse, S. Wagner, D. Pfefferer, R. Lawaczeck, V. Elste, W. Semmler, Magn. Reson. Med. **40**, 236–242 (1998)
89. C.L. Dennis, K.L. Krycka, J.A. Borchers, R.D. Desautels, J. van Lierop, N.F. Huls, A.J. Jackson, C. Gruettner, R. Ivkov, Adv. Func. Mater. **25**, 4300–4311 (2015)
90. C. Blanco-Andujar, D. Ortega, P. Southern, Q.A. Pankhurst, N.T.K. Thanh, Nanoscale **7**, 1768–1775 (2015)
91. B. Luigjes, S.M.C. Woudenberg, R. de Groot, J.D. Meeldijk, H.M. Torres Galvis, K.P. de Jong, A.P. Philipse, B.H. Erné, J. Phys. Chem. C. **115**, 14598–14605 (2011)
92. GLOBOCAN 2020, Cancer incidence, https://gco.iarc.fr/today/. Accessed 14 April 2023
93. A.K. Abbas, A.H. Lichtman, Basic immunology: functions and disorders of the immune system—NLM Catalog—NCBI, United State, 3rd ed. (2015)
94. L.Q. Chow, C. Chen, D. Raben, Curr. Cancer Therapy Rev. **3**, 255–266 (2007)
95. B.P. Pollack, B. Sapkota, T.V. Cartee, Clin. Cancer Res. **17**, 4400–4413 (2011)
96. K.C. Valkenburg, A.E. de Groot, K.J. Pienta, Nat. Rev. Clin. Oncol. **15**, 366–381 (2018)
97. K. Thankappan, S. Iyer, J. Menon, *Dysphagia Management in Head and Neck Cancers A Manual and Atlas: A Manual and Atlas* (2018)
98. X. Gao, S. Wang, Z. Tian, Y. Wu, W. Liu, Transl. Cancer Res. **10**, 251–260 (2021)
99. D. Pulte, H. Brenner, Oncologist **15**, 994–1001 (2010)
100. S. Indoria, V. Singh, M.-F. Hsieh, Int. J. Pharm. **582**, 119314 (2020)
101. A.-G. Niculescu, A.M. Grumezescu, Int. J. Mol. Sci. **23**, 5253 (2022)
102. A. Aghebati-Maleki, S. Dolati, M. Ahmadi, A. Baghbanzhadeh, M. Asadi, A. Fotouhi, M. Yousefi, L. Aghebati-Maleki, J. Cell. Physiol. **235**, 1962–1972 (2020)
103. M.J. Mitchell, M.M. Billingsley, R.M. Haley, M.E. Wechsler, N.A. Peppas, R. Langer, Nat. Rev. Drug Discov. **20**, 101–124 (2021)
104. M.F. Attia, N. Anton, J. Wallyn, Z. Omran, T.F. Vandamme, J. Pharm. Pharmacol. **71**, 1185–1198 (2019)
105. A. Ruiz, G. Salas, M. Calero, Y. Hernández, A. Villanueva, F. Herranz, S. Veintemillas-Verdaguer, E. Martínez, D.F. Barber, M.P. Morales, Acta Biomater. **9**, 6421–6430 (2013)
106. N. Hoshyar, S. Gray, H. Han, G. Bao, Nanomedicine (Lond.) **11**, 673–692 (2016)
107. I. Hilger, Int. J. Hyperth. **29**, 828–834 (2013)
108. X. Liu, Y. Zhang, Y. Wang, W. Zhu, G. Li, X. Ma, Y. Zhang, S. Chen, S. Tiwari, K. Shi, Theranostics **10**, 3793 (2020)
109. D. Chang, M. Lim, J.A. Goos, R. Qiao, Y.Y. Ng, F.M. Mansfeld, M. Jackson, T.P. Davis, M. Kavallaris, Front. Pharmacol. **9**, 831 (2018)
110. O.S. Nielsen, M. Horsman, J. Overgaard, Eur. J. Cancer **37**, 1587–1589 (2001)
111. A. Villanueva, P. de la Presa, J.M. Alonso, T. Rueda, A. Martínez, P. Crespo, M.P. Morales, M.A. Gonzalez-Fernandez, J. Valdés, G. Rivero, J. Phys. Chem. C **114**, 1976–1981 (2010)
112. G.F. Goya, L. Asín, M.R. Ibarra, Int. J. Hyperth. **29**, 810–818 (2013)
113. J.R. Lepock, Int. J. Hyperth. **21**, 681–687 (2005)
114. J.R. Lepock, Methods **35**, 117–125 (2005)
115. A.J. Peer, M.J. Grimm, E.R. Zynda, E.A. Repasky, Immunol. Res. **46**, 137–154 (2010)
116. P. de Andrade Mello, S. Bian, L.E.B. Savio, H. Zhang, J. Zhang, W. Junger, M. R. Wink, G. Lenz, A. Buffon, Y. Wu, S.C. Robson, Oncotarget **8**, 67254–67268 (2017)
117. K.L. Eales, K.E.R. Hollinshead, D.A. Tennant, Oncogenesis **5**, e190 (2016)
118. D. Ortega, Q.A. Pankhurst, in *Nanoscience,* vol. 1, ed. by P. O'Brien (Royal Society of Chemistry, Cambridge, 2012), pp. 60–88
119. R. Hergt, S. Dutz, J. Magn. Magn. Mater. **311**, 187–192 (2007)
120. M. Domenech, I. Marrero-Berrios, M. Torres-Lugo, C. Rinaldi, ACS Nano **7**, 5091–5101 (2013)
121. M. Creixell, A.C. Bohórquez, M. Torres-Lugo, C. Rinaldi, ACS Nano **5**, 7124–7129 (2011)
122. H.F. Krug, P. Wick, Angew. Chem. Int. Ed. Engl. **50**, 1260–1278 (2011)
123. A. Elengoe, S. Hamdan, Int. J. Advanc. Life Sci. Res. 22–27 (2018)

124. I. Takahashi, Y. Emi, S. Hasuda, Y. Kakeji, Y. Maehara, K. Sugimachi, Surgery **131**, S78–S84 (2002)
125. Z. Behrouzkia, Z. Joveini, B. Keshavarzi, N. Eyvazzadeh, R.Z. Aghdam, Oman Med. J. **31**, 89–97 (2016)
126. V. Ejigah, O. Owoseni, P. Bataille-Backer, O.D. Ogundipe, F.A. Fisusi, S.K. Adesina, Polymers **14**, 2601 (2022)
127. Y. Matsumura, H. Maeda, Cancer Res. **46**, 6387–6392 (1986)
128. E. Alphandéry, S. Faure, L. Raison, E. Duguet, P.A. Howse, D.A. Bazylinski, J. Phys. Chem. C **115**, 18–22 (2011)
129. M.K. Yu, J. Park, S. Jon, Theranostics **2**, 3–44 (2012)
130. S. Wilhelm, A.J. Tavares, Q. Dai, S. Ohta, J. Audet, H.F. Dvorak, W.C.W. Chan, Nat. Rev. Mater. **1**, 1–12 (2016)
131. R. Bazak, M. Houri, S. El Achy, S. Kamel, T. Refaat, J. Cancer Res. Clin. Oncol. **141**, 769–784 (2015)
132. A.L.C. de S.L. Oliveira, T. Schomann, L.-F. de Geus-Oei, E. Kapiteijn, L.J. Cruz, R.F. de Araújo Junior, Pharmaceutics **13**, 1321 (2021)
133. E.Y. Makhani, A. Zhang, J.B. Haun, Nano Convergence **8**, 38 (2021)
134. V. Torchilin, Adv. Drug Deliv. Rev. **63**, 131–135 (2011)
135. S. Chono, T. Tanino, T. Seki, K. Morimoto, J. Pharm. Pharmacol. **59**, 75–80 (2007)
136. H. Maeda, T. Sawa, T. Konno, J. Control. Release **74**, 47–61 (2001)
137. H. Kobayashi, R. Watanabe, P.L. Choyke, Theranostics **4**, 81–89 (2013)
138. M.O. Oyewumi, R.A. Yokel, M. Jay, T. Coakley, R.J. Mumper, J. Control. Release **95**, 613–626 (2004)
139. A. Angelopoulou, A. Kolokithas-Ntoukas, C. Fytas, K. Avgoustakis, ACS Omega **4**, 22214–22227 (2019)
140. M. Truffi, M. Colombo, L. Sorrentino, L. Pandolfi, S. Mazzucchelli, F. Pappalardo, C. Pacini, R. Allevi, A. Bonizzi, F. Corsi, D. Prosperi, Sci. Rep. **8**, 6563 (2018)
141. J. Yoo, C. Park, G. Yi, D. Lee, H. Koo, Cancers (Basel) **11**, E640 (2019)
142. G.T. Tietjen, L.G. Bracaglia, W.M. Saltzman, J.S. Pober, Trends Mol. Med. **24**, 598–606 (2018)
143. C. NDong, J.A. Tate, W.C. Kett, J. Batra, E. Demidenko, L.D. Lewis, P.J. Hoopes, T.U. Gerngross, K.E. Griswold, PLoS One **10**, e0115636 (2015)
144. S. Palanisamy, Y.-M. Wang, Dalton Trans. **48**, 9490–9515 (2019)
145. V. Mittelheisser, P. Coliat, E. Moeglin, L. Goepp, J.G. Goetz, L.J. Charbonnière, X. Pivot, A. Detappe, Adv. Mater. **34**, 2110305 (2022)
146. E. Alphandéry, Nanotoxicology **13**, 573–596 (2019)
147. A. Ahmad, F. Khan, R.K. Mishra, R. Khan, J. Med. Chem. **62**, 10475–10496 (2019)
148. J.D. Byrne, T. Betancourt, L. Brannon-Peppas, Adv. Drug Deliv. Rev. **60**, 1615–1626 (2008)
149. T.M. Allen, Nat. Rev. Cancer **2**, 750–763 (2002)
150. G.T. Hermanson, *Bioconjugate Techniques* (Academic Press, 2013)
151. T. Iwasawa, P. Wash, C. Gibson, J. Rebek, Tetrahedron **63**, 6506–6511 (2007)
152. D.P. Nair, M. Podgórski, S. Chatani, T. Gong, W. Xi, C.R. Fenoli, C.N. Bowman, Chem. Mater. **26**, 724–744 (2014)
153. N. Guldris, J. Gallo, L. García-Hevia, J. Rivas, M. Bañobre-López, L.M. Salonen, Chem. Euro. J. **24**, 8624–8631 (2018)
154. J. Pellico, P.J. Gawne, R.T.M. de Rosales, Chem. Soc. Rev. **50**, 3355–3423 (2021)
155. J. Nam, N. Won, J. Bang, H. Jin, J. Park, S. Jung, S. Jung, Y. Park, S. Kim, Adv. Drug Deliv. Rev. **65**, 622–648 (2013)
156. B. Freis, M.D.L.Á. Ramírez, S. Furgiuele, F. Journe, C. Cheignon, L.J. Charbonnière, C. Henoumont, C. Kiefer, D. Mertz, C. Affolter-Zbaraszczuk, F. Meyer, S. Saussez, S. Laurent, M. Tasso, S. Bégin-Colin, Int. J. Pharm. **635**, 122654 (2023)
157. G. Rimkus, S. Bremer-Streck, C. Grüttner, W.A. Kaiser, I. Hilger, Contrast Media Mol. Imag. **6**, 119–125 (2011)
158. P. Foroozandeh, A.A. Aziz, Nanoscale Res. Lett. **13**, 339 (2018)

159. B. Svitkova, V. Zavisova, V. Nemethova, M. Koneracka, M. Kretova, F. Razga, M. Ursinyova, A. Gabelova, Beilstein J. Nanotechnol. **12**, 270–281 (2021)
160. H. Hossein-Nejad-Ariani, E. Althagafi, K. Kaur, Sci. Rep. **9**, 2723 (2019)
161. M.W. Freeman, A. Arrott, J.H.L. Watson, J. Appl. Phys. **31**, S404–S405 (1960)
162. Y. Shen, X. Liu, Z. Huang, N. Pei, J. Xu, Z. Li, Y. Wang, J. Qian, J. Ge, Cell Transplant. **24**, 1981–1997 (2015)
163. S. Zhao, X. Yu, Y. Qian, W. Chen, J. Shen, Theranostics **10**, 6278–6309 (2020)
164. R. Baskar, K.A. Lee, R. Yeo, K.-W. Yeoh, Int. J. Med. Sci. **9**, 193–199 (2012)
165. Y. Liu, P. Zhang, F. Li, X. Jin, J. Li, W. Chen, Q. Li, Theranostics **8**, 1824–1849 (2018)
166. K. Haume, S. Rosa, S. Grellet, M.A. Śmiałek, K.T. Butterworth, A.V. Solov'yov, K.M. Prise, J. Golding, N.J. Mason, Cancer Nanotechnol. **7**, 8 (2016)
167. J.F. Hainfeld, D.N. Slatkin, H.M. Smilowitz, Phys. Med. Biol. **49**, N309-315 (2004)
168. S. Li, S. Bouchy, S. Penninckx, R. Marega, O. Fichera, B. Gallez, O. Feron, P. Martinive, A.-C. Heuskin, C. Michiels, S. Lucas, Nanomedicine (London) **14**, 317–333 (2019)
169. N.G. Shetake, A. Kumar, B.N. Pandey, Biochimica et Biophysica Acta (BBA)—General Subjects **1863**, 857–869 (2019)
170. R. Ahmad, G. Schettino, G. Royle, M. Barry, Q.A. Pankhurst, O. Tillement, B. Russell, K. Ricketts, Part. Part. Syst. Charact. **37**, 1900411 (2020)
171. I. Ternad, S. Penninckx, V. Lecomte, T. Vangijzegem, L. Conrard, S. Lucas, A.-C. Heuskin, C. Michiels, R.N. Muller, D. Stanicki, S. Laurent, Nanomaterials **13**, 201 (2023)
172. V. Schirrmacher, Int. J. Oncol. **54**, 407–419 (2019)
173. A. Mohammad, Open Access J. Toxicol. https://doi.org/10.19080/OAJT.2018.02.555600
174. D. Huang, H. Duan, H. Huang, X. Tong, Y. Han, G. Ru, L. Qu, C. Shou, Z. Zhao, Sci. Rep. **6**, 20502 (2016)
175. K.B. Challagundla, P.M. Wise, P. Neviani, H. Chava, M. Murtadha, T. Xu, R. Kennedy, C. Ivan, X. Zhang, I. Vannini, F. Fanini, D. Amadori, G.A. Calin, M. Hadjidaniel, H. Shimada, A. Jong, R.C. Seeger, S. Asgharzadeh, A. Goldkorn, M. Fabbri, JNCI: J. Nat. Cancer Instit. **107**, djv135 (2015)
176. X. Wang, H. Zhang, X. Chen, Cancer Drug Resistance **2**, 141–160 (2019)
177. Y. Huang, K. Mao, B. Zhang, Y. Zhao, Mater. Sci. Eng. C **70**, 763–771 (2017)
178. P. Eslami, M. Albino, F. Scavone, F. Chiellini, A. Morelli, G. Baldi, L. Cappiello, S. Doumett, G. Lorenzi, C. Ravagli, A. Caneschi, A. Laurenzana, C. Sangregorio, Nanomaterials (Basel) **12**, 303 (2022)
179. M.H. El-Dakdouki, D.C. Zhu, K. El-Boubbou, M. Kamat, J. Chen, W. Li, X. Huang, Biomacromol **13**, 1144–1151 (2012)
180. M.H. El-Dakdouki, J. Xia, D.C. Zhu, H. Kavunja, J. Grieshaber, S. O'Reilly, J.J. McCormick, X. Huang, A.C.S. Appl, Mater. Interfaces **6**, 697–705 (2014)
181. S.-W. Lee, Y.-M. Kim, C.H. Cho, Y.T. Kim, S.M. Kim, S.Y. Hur, J.-H. Kim, B.-G. Kim, S.-C. Kim, H.-S. Ryu, S.B. Kang, Cancer Res. Treat. **50**, 195–203 (2018)
182. E. Guisasola, L. Asín, L. Beola, J.M. de la Fuente, A. Baeza, M. Vallet-Regí, A.C.S. Appl, Mater. Interfaces **10**, 12518–12525 (2018)
183. Y. Fang, Y. He, C. Wu, M. Zhang, Z. Gu, J. Zhang, E. Liu, Q. Xu, A.M. Asrorov, Y. Huang, Theranostics **11**, 6860–6872 (2021)
184. T. Vangijzegem, V. Lecomte, I. Ternad, L. Van Leuven, R.N. Muller, D. Stanicki, S. Laurent, Pharmaceutics **15**, 236 (2023)
185. B. Li, L. Lin, Light Sci Appl **11**, 85 (2022)
186. T.-J. Yu, P.-H. Li, T.-W. Tseng, Y.-C. Chen, Nanomedicine **6**, 1353–1363 (2011)
187. H. Chen, J. Burnett, F. Zhang, J. Zhang, H. Paholak, D. Sun, J. Mater. Chem. B **2**, 757–765 (2014)
188. M.-Y. Liao, P.-S. Lai, H.-P. Yu, H.-P. Lin, C.-C. Huang, Chem. Commun. **48**, 5319 (2012)
189. S. Shen, F. Kong, X. Guo, L. Wu, H. Shen, M. Xie, X. Wang, Y. Jin, Y. Ge, Nanoscale **5**, 8056 (2013)
190. Z. Zhou, Y. Sun, J. Shen, J. Wei, C. Yu, B. Kong, W. Liu, H. Yang, S. Yang, W. Wang, Biomaterials **35**, 7470–7478 (2014)

191. E.A. Hussein, M.M. Zagho, B.R. Rizeq, N.N. Younes, G. Pintus, K.A. Mahmoud, G.K. Nasrallah, A.A. Elzatahry, Int. J. Nanomed. **14**, 4529–4539 (2019)
192. A. Pinto, M. Pocard, Pleura and Peritoneum. https://doi.org/10.1515/pp-2018-0124
193. L. Wu, L. Chen, F. Liu, X. Qi, Y. Ge, S. Shen, Colloids Surf. B **152**, 440–448 (2017)
194. X. Sun, B. Liu, X. Chen, H. Lin, Y. Peng, Y. Li, H. Zheng, Y. Xu, X. Ou, S. Yan, Z. Wu, S. Deng, L. Zhang, P. Zhao, J. Mater. Sci. Mater. Med. **30**, 76 (2019)
195. J. R. Kanwar, S. K. Kamalapuram, S. Krishnakumar, R. K. Kanwar, Nanomedicine **11**(3), 249–268 (2016). https://doi.org/10.2217/nnm.15.199
196. C.-C. Wu, Y.-C. Yang, Y.-T. Hsu, T.-C. Wu, C.-F. Hung, J.-T. Huang, C.-L. Chang, Oncotarget **6**, 26861–26875 (2015)